THE BOOK OF SQUARES

Incipit liber quadrato[illegible] [illegible] a leonardo pis. Anni. M. cc. xxv.
CUM . Magister dominicus pedib[illegible] celsitudinis vre princeps
gloriosissime dne. F. me pis. duceret [illegible] occurrens
Magr Johanes panormitanus [illegible] mi [illegible] infra
scriptam non minus ad geometriam q ad [illegible] pertinet . ut invenire
num quadratu cu quo [illegible] additis ut diminutis sep in qdrat
nus oriret . sup cui quois solutione a me iam inventa [illegible]
unde q [illegible] origine solutio ipa [illegible] qdratis & int
qdratos nuos accidit . Nu[illegible] aut cu relatoib[illegible] pis [illegible] &
alio[illegible] redeuntu[illegible] [illegible] curia [illegible] qd [illegible] sub-
limitas maiestas legit [illegible] libru que [illegible] de nuo & q
placet uobis audire aliquoties subtilitates ad geometria
& nuri [illegible] [illegible] curia & [illegible] pho
[illegible] in [illegible] qoem . [illegible] sup[illegible] [illegible]
ad uram honore condere [illegible] qd uocari libru uolui
quadratoz . [illegible] postulas patienter . [illegible] plus ut
[illegible] iusto sit [illegible] continet . cum oiu[illegible] [illegible] memoria
& in nullo peccare sit diuinitatis potius q humanitatis et
nemo sit uitio carens & undiq circumspectus .

CONSIDERAVI sup origine oiu[illegible] qdratoz nuoz & [illegible] ipaz [illegible]
ex ordinata iparium ascensioe . Na unitas qdrata e & ex ipa
efficit primus qdrat . s. unu cui unitati addito [illegible] . facit sm
qdratu . s. 4 cui radix e 2 . cui et additioi si addat t[illegible] ipar-
nus . s. 5 ttius qdratus [illegible] . s. 9 cui radix e 3 . & sic sep porcdi-
nata iparium collectione . ordinata [illegible] et series qdratoz
vn cu uolui . ii . quadratos nuos [illegible] quoz additio faciat
qdratu . [illegible] accipia quale uoluero qdratu ipare . & [illegible]
ipm p uno ex duob dictis quadratis . reliqum inuenia ex collectione oiu[illegible] iparium q st ab unitate [illegible] ad ipm qdratu
ipare . Vbi gra accipia 9 p uno ex dictis duob qdratis reliqu
hebo ex collectioe oiu[illegible] iparium q st sub . 9 . s. de 1 et 3 et 5 et 7 .
q[illegible] summa e 16 q e qdrat q addito cu . 9 . [illegible] 25 . q nus est
qdrat . et si geometrice uti uoluim demonstratioe . ladiuetat quot
cuq nun ipares ab unitate p ordine ascendendo donec extrem
eoz qdratus fiat & sint . ab . bc . cd . de . ef . & sit . ef . qdratus
ergo . ef . e qdrat & ae e quadrat cu [illegible] ex ordinata colle-
tione iparium . ab . et bc . et cd . & de . et totus . af . nus e simile
quadrat et sic ex duob qdratis ae & ef . fit qdratus . af .
Item aliter accipia aliqe quadratu pare cui medietas sit par

a b c d e f

Beginning of *Liber quadratorum* by Leonardo Pisano. From "Il Libro dei Quadrati di Leonardo Pisano" in *Physis: Revista Internazionale di Storia della Scienza*, 1979, p. 197. Reprinted with permission of the author, Ettore Picutti.

LEONARDO PISANO
FIBONACCI

THE BOOK OF SQUARES

An Annotated Translation into
Modern English by L.E. Sigler

1987

ACADEMIC PRESS, INC.
Harcourt Brace Jovanovich, Publishers
Boston Orlando San Diego
New York Austin London Sydney
Tokyo Toronto

ACADEMIC PRESS, INC.
Orlando, Florida 32887

Library of Congress Cataloging-in-Publication Data

Fibonacci, Leonardo, ca. 1170–ca. 1240.
The book of squares.

Translation of: Liber quadratorum.
Bibliography: p.
Includes indexes.
1. Mathematics—Early works to 1800. 2. Numbers,
Theory of—Early works to 1800. 3. Fibonacci, Leonardo,
ca. 1170—ca. 1240. I. Sigler, L. E. II. Title.
QA32.F4813 1986 510 86-17336
ISBN 0-12-643130-2 (alk. paper)

9 8 7 6 5 4 3 2 1
Printed in USA

CONTENTS

PREFACE

Ancient mathematical works, when translated into modern language and notation, can make very good reading for contemporary students of mathematics. Far from being irrelevant and obsolete, they contain valid and interesting mathematics, historically interesting science. A student of mathematics can still enjoy reading the work of a master; such a work is *Liber quadratorum*, written in 1225 by Leonardo Pisano (Leonardo of Pisa, usually known as Fibonacci by mathematicians). This volume has three main parts: a short biographical sketch of Leonardo Pisano, an English translation of *Liber quadratorum*, and a detailed commentary on the translation, *The Book of Squares*.

The biographical sketch provides details of Leonardo's life and an outline of his works. *The Book of Squares* is a translation into English from the Latin text prepared and published in 1862 by Baldassarre Boncompagni [B], who found the manuscript in the Ambrosian Library in Milan where it had lain unknown for many years. The commentary contains a step-by-step explanation of Leonardo's work in modern mathematical notation and terminology. It is intended to make it easier for the reader to follow *The Book of Squares.*

The publication of this translation of *Liber quadratorum* (*The Book of Squares*) is to make available in English for mathematics students of all ages and the general public an important work of Leonardo Pisano, the greatest by far of Western mathematicians of the Middle Ages. A knowledge of secondary school mathematics, algebra and geometry ought to be adequate preparation for the reading and understanding of this book. *Liber quadratorum* was translated from Latin to French in 1952 by Paul Ver Eecke [Ve]. When I first read Leonardo's work, it immediately struck me how much American students would enjoy reading this delightful mathematics. I obtained the Latin edition by Boncompagni and translated the work into English. I have since learnt that a significant portion of the work was published in English by Mr. Edward Grant in 1974 in *A Source Book in Medieval Science* [Gr]. I believe that the interests of students of mathematics will be better served by having available a complete version of *The Book of Squares*, which can be read and enjoyed primarily as a book of mathematics. This English translation is very literal to follow Leonardo's words as closely as possible; although some may find the language awkward at times, I hope thereby to reproduce Leonardo's thoughts more faithfully as well as to give an impression of the age of the work. The English translation of each section is followed by the commentary. This is intended to assist the reader in following the arguments in *The Book of Squares.* The commentary generally contains the mathematical argument in modern notation with some assistance in bridging the transition. The reader who finds some difficulty in following the text directly may find it helps to read the commentary first.

Some will find it most useful to go back and forth between the text and the commentary. I have chosen to put all hints and remarks in the commentary rather than to use footnotes in order to preserve the continuity of the text.

The geometrical algebra used by Leonardo is that presented by Euclid in the *Elements*: Leonardo had great facility with geometrical algebra. Rather than discuss and refer to the *Elements* every detail of geometrical algebra, I have concentrated instead on helping the reader follow the argument. A reader who understands the mathematical argument being presented will readily understand Leonardo's notation. Leonardo's ideas and arguments are elegant and they deserve the utmost respect; I have attempted to render them faithfully. It is easy to be impressed by the elegance and intricacy of his arguments in a geometrical notation that certainly did not facilitate algebraic insight. Yet Leonardo's mastery and dexterity with the crude algebra are astonishing.

The Book of Squares explores the relation of square numbers to sums of sequences of odd numbers. Leonardo takes this simple relation and builds an impressive amount of mathematical theory and results. He ingeniously solves a quantity of problems building on the properties of squares as sums of odd numbers. The problems are in indeterminate algebra of the kind studied by Diophantus (250 A.D.) [H1], but Leonardo uses methods for solution which are his own. It is known that Leonardo learnt much mathematics from Arabic sources in Bugia (North Africa) and other Islamic lands. One cannot know for certain that *The Book of Squares* is completely original when one knows so little of the history of mathematics of the relevant periods of time. Franz Woepcke in his translation of al-Kharkhi's (also spelled al-Karaji) *Fakhri* [Wo, pp. 24–31, 143–147] points out similarities and differences between the *Fakhri* and Leonardo's work. Ver Eecke [Ve] went so far as to accuse

Leonardo of borrowing numerical values from Diophantus' *Arithmetica* with no adequate motivation. That the values do occur in *Arithmetica* is true, but it is false that Leonardo does not offer motivation. Mr. Ettore Picutti [P2, p. 39] ably defends Leonardo from these false criticisms and argues convincingly for Leonardo's originality. In his most erudite study Mr. Picutti establishes Leonardo as the founder of the Tuscan school of mathematicians, putting an end to the superficial appearance of Leonardo's being isolated in time and place. It is known that Diophantus' work was studied thoroughly by mathematicians writing in Arabic (parts of the *Fakhri* were apparently copied from the *Arithmetica* of Diophantus [S, pp. 10–11]) who were interested in the problems that Leonardo treated. Mr. A. Anbouba has noted similarities between the work of Leonardo and the work of al-Khazin [A, p. 149]. Mr. Jacques Sesiano has speculated in his translation of the Arabic version of *Arithmetica* that both *The Book of Squares* and the al-Khazin work may have some earlier common source [S, p. 83]. As study continues on the history of mathematics, especially of this period, more details on the works of the mathematicians writing in Arabic will likely become available. It is clear, however, that Leonardo had a magnificent command of the mathematics he presented and was an accomplished mathematician. Although it has been shown that there are antecendents for the problems he considered, and that he must have used material from Arabic language sources, no one has shown that his methods and work were not original. The freshness and quality of his work insist that Leonardo Pisano was a great and original creative mathematician and well deserves being called the greatest Western mathematician of the Middle Ages.

The book as written by Leonardo is continuous without numbering of theorems. R. B. McClenon numbered the theorems in his analysis [Mc] and so did Ver Eecke, in his French translation. I have also done this but the numberings are not the same as those of McClenon nor of Ver Eecke. The notation for line segments is that found in Boncompagni's Latin edition. I apologize to the reader for the inevitable errors in translation and copying as well as for any errors of judgment and perception.

The author wishes to acknowledge gratefully the generous assistance given by Mr. André Weil for the publication of this book.

INTRODUCTION
A Brief Biography of Leonardo Pisano (Fibonacci) [1170–post 1240]

Leonardo Pisano was born in 1170 in the city-state of Pisa in the province of Tuscany in what is now the state of Italy. His father's name was Guilielmo and Leonardo also identified himself as a descendant of Bonaccio, most probably a not far removed progenitor. This reference to a famous ancestor was then a common practice in Italy. In 1225, in his book *Flos*, Leonardo refers to himself as Leonardo Pisano Bigollo, and later in 1240 in an official document of Pisa awarding him an honorarium for service as a financial advisor he is referred to as Leonardo Pisano Bigollo [L, p. 5].

Many unsuccessful attempts have been made to explain the meaning of *Bigollo*, some quite fanciful, but there is no merit in repeating them here. It is, however, quite worth saying that the use of the sobriquet *Fibonacci* for Leonardo Pisano probably originated with the mathematical historian Guillaume Libri in 1838; there is no evidence that Leonardo so referred to himself or was ever so called by his contemporaries [P2, p. 36]. Nevertheless, as has been often the case with mathematical history, these lamentable errors or fantasies catch on, persist, and seem never to be correctable. Leonardo Pisano is more well known today among mathematicians as Fibonacci than by his real name.

Later in his life, after returning to Pisa from his work and travels, Leonardo became associated with the court of Frederick II, emperor of the Holy Roman Empire. Pisa was during this era an independent city-state of the Italian peninsula, but part of the Holy Roman Empire. The empire was ruled by Frederick II, a German prince from Swabia, who made Palermo in Sicily the principal capital of his empire. Frederick was called the wonder of the world in tribute for the scholarship and splendor of his court and for his own learning and magnificence. Frederick fitted in every way the criteria for the enlightened ruler and prince. When Frederick traveled over the empire he was generally accompanied by his musicians, poets, scholars, philosophers, and pets. Frederick II took a genuine personal as well as imperial interest in the learned activities of his court and was himself the author of a book on falconry [K].

In the beginning of his popular book *Liber abbaci*, dedicated to the mediaeval scholar Michael Scott, Leonardo gives us a short autobiography.

"I joined my father after his assignment by his homeland Pisa as an officer in the customhouse located at Bugia [Algeria] for the Pisan merchants who were often there. He had me marvelously instructed in the Arabic-Hindu numerals and calculation. I enjoyed so much the instruction that I later continued to study mathematics while on business trips to Egypt, Syria, Greece, Sicily, and Provence and there enjoyed discussions and disputations with the scholars of those places. Returning to Pisa I composed this book of fifteen chapters which comprises what I feel is the best of the Hindu, Arabic, and Greek methods. I have included proofs to further the understanding of the reader and the

> Italian people. If by chance I have omitted anything more or less proper or necessary, I beg forgiveness, since there is no one who is without fault and circumspect in all matters."

Knowledge of Leonardo's late life and death is unfortunately nonexistent.

In addition to *Liber abbaci* (1202, 1228), his book on mathematics and calculation, Leonardo Pisano wrote *Practica geometriae* (1223), *Flos* (1225), *Epsistola ad Magistrum Theodorum* (?), *Liber quadratorum* (1225), and now lost, a book on commercial arithmetic, *Di minor guisa*, and possibly a tract on Book X of the *Elements.* Leonardo belonged to or was associated with the scholars of the imperial court. Leonardo mentions Michael Scott, Master Theodore, and Master John of Palermo by name. Leonardo's friend, Master Dominick, presented him to the court when it was held in Pisa about 1225.

In the title *Liber abbaci*, abbaci has the more general meaning of mathematics and calculation or applied mathematics rather than merely of the counting machine made from stringing beads on wires. The mathematicians of Tuscany following Leonardo were called Maestri d'Abbaco; for more than three centuries there were masters and students trained in mathematics and calculation based on the principles established in *Liber abbaci* [Vo, p. 612]. So also were trained the *Cossists* and *Rechenmeister* of Germany in his tradition. Notable in the Tuscan school were Paolo dell'Abaco, Cristofano di Gherardo, Master Benedetto, and Luca Pacioli. Master Benedetto wrote an Italian version of *Liber quadratorum* in 1464. This work in Italian has been published in a most careful and admirable study by Mr. Ettore Picutti [P1]. Brother Luca Pacioli's book *Summa*, printed first in Venice in 1494 and reprinted in 1523, lauded Leonardo's work and copied at length passages and problems from *Liber quadratorum.*

Leonardo's stated purpose in writing *Liber abbaci* was to introduce Arabic numerals and methods of arithmetic into Italy. He deemed them vastly superior to Roman numerals then in use in business and accounting. Even in the Middle Eastern Muslim lands, Arabic numerals and calculation were used only by mathematicians and scientists in their scientific works [Go, p. 209] and not by businessmen and accountants. Leonardo did not introduce Arabic business calculations into Italy, but can rather be credited with the greater innovation of introducing scientific calculating methods into general business practice. The historian Mr. S. Goitein points out that it was the Europeans who eventually taught Arab businessmen the superiority of Arabic numerals and calculation for business and accounting [Go, p. 209].

A partial list of contents for *Liber abbaci* includes these subjects: the nine Arabic numerals (ten counting zero); calculation with these numerals; multiplication and addition of whole numbers; subtraction of whole numbers from larger numbers; division of whole numbers; addition, multiplication, and division by fractions; decomposition of whole numbers into parts; barter, exchange and rules for money; miscellaneous problems; method of double false position; quadratic and cubic root extraction; analysis of quadratic equations, binomials, rules of proportion, algebraic rules, casting out nines, accounting, progressions; many problems in applied algebra including problems in indeterminate analysis. Included is the famous rabbit problem, which leads to the sequence named Fibonacci by Edouard Lucas, who studied it. The sequence was not studied by Leonardo. Included in *Liber abbaci* are arguments for the validity of the methods. Although his purpose was to disseminate methods for practical use, Leonardo remained very much the mathematician; he gave arguments and proofs in Euclidean tradition. He used general letters to represent unknown quantities. Although he does not use modern logical equations employing symbols for operations, the sentences of his works are directly translatable into modern equations. One can wonder what mental visual schemes Leonardo used when he worked with his algebra.

Encouraged by his friend, Master Dominick, Leonardo Pisano published *Practica geometriae* in 1223. The book deals with a number of geometrical subjects and is based upon both the geometry of Euclid and that of Heron of Alexandria. Sections of the work are borrowed from *Liber embadorum* (1145) of Plato of Tivoli, which was based upon Arabic works [Vo, p. 609] and ultimately goes back to Euclid's lost book *Division of Figures.* Included in Leonardo's geometry are many practical problems on surveying and land measurement. Included also is the famous formula of Heron for the area of a triangle in terms of the sides. The last section of the book is devoted to solution of indeterminate equations having little to do with practical geometry. This interest in indeterminate equations comes to greatest fruition in *Liber quadratorum.*

Such interest in solution of equations is found in his works *Flos* (*Flower*) (1225) and *Epistola ad Magistrum Theodorum* (*Letter to Master Theodore*) (?). Besides indeterminate analysis, he also finds a very accurate approximation to the real root of a cubic equation. Leonardo distinguishes between integral solutions and rational solutions to indeterminate equations. The term *Diophantine analysis* might more properly be called *Leonardine analysis.*

Liber quadratorum was written in 1225 and dedicated to the emperor. It is his most advanced book and represents his greatest achievement as a mathematician. From the *Arithmetica* of Diophantus to the work of Fermat, it is the primary work on arithmetic (today called the theory of numbers).

Pacioli quoted extensively from *Liber quadratorum* in his book *Summa* published in 1494, but *Liber quadratorum* was never published during the Renaissance. The work of Leonardo Pisano was known to Tartaglia, but fell into oblivion until the end of the eighteenth century when the historian Pietro Cossali, intrigued by Pacioli's references, searched in vain for a manuscript of Leonardo's *Liber quadratorum.* He then made an attempt to reconstruct the book from the data in *Summa.* A manuscript was eventually found in the middle of the nineteenth century by the great mediaeval scholar Baldassarre Boncompagni in the Ambrosian Library in Milan. Other manuscripts have since appeared, including the Italian version by Master Benedetto; this English translation is from the Latin text prepared and corrected by Boncompagni from the manuscript in Milano.

The mathematical work of Leonardo Pisano was innovative, creative, and high in quality. He introduced into Europe the numerals, calculation, and algebra of the Orient. Yet he was, after all, a profound scholar of Greek mathematics, a worthy successor of Euclid, Archimedes, Heron, and Diophantus. He stands above all other mathematicians from classical times to the Renaissance. He joined the theoretical tradition of the Hellenes and the algebraic tradition of the Arabs and established them in Europe. Leonardo Pisano is the great European mathematician of the Middle Ages.

THE BOOK OF SQUARES

Here begins the Book of Squares composed by Leonardo Pisano in the year 1225.

Prologue

After being brought to Pisa by Master Dominick to the feet of your celestial majesty, most glorious prince, Lord F., I met Master John of Palermo; he proposed to me a question that had occurred to him, pertaining not less to geometry than to arithmetic: find a square number from which, when five is added or subtracted, always arises a square number. Beyond this question, the solution of which I have already found, I saw, upon reflection, that this solution itself and many others have origin in the squares and the numbers which fall between the squares. When I heard recently from a report from Pisa and another from the Imperial Court that your sublime majesty deigned to read the book I composed on numbers, and that it pleased you to listen to several subtleties touching on geometry and numbers, I recalled the question proposed to me at your court by your philosopher. I took upon myself the subject matter and began to compose in your honor this work which I wish to call *The Book of Squares*. I have come to request indulgence if in any place it contains something more or less than right or necessary; for to remember everything and be mistaken in nothing is divine rather than human; and no one is exempt from fault nor is everywhere circumspect.

Comments on the Prologue

Leonardo speaks of his introduction to Frederick II (1194–1250), emperor of the Holy Roman Empire, an extraordinary patron of learning, science, and the arts. Frederick maintained at his court a large group of learned men including the mentioned John of Palermo. Pisa in

1225 was a prosperous naval republic which formed part of the empire and supported Frederick II in his efforts against numerous plots and rebellions in the Italian peninsula. The previous work of Leonardo to which he refers most likely is *Liber abbaci,* written in 1202. The problem referred to by Leonardo—to find a square number from which, when five is added or subtracted, always yields a square number—is proposition 17 of this work. The problem itself is in the tradition of Diophantine indeterminate algebra. Most probably, John of Palermo obtained the problem from Arabic sources.

Introduction

I thought about the origin of all square numbers and discovered that they arise out of the increasing sequence of odd numbers; for the unity is a square and from it is made the first square, namely 1; to this unity is added 3, making the second square, namely 4, with root 2; if to the sum is added the third odd number, namely 5, the third square is created, namely 9, with root 3; and thus sums of consecutive odd numbers and a sequence of squares always arise together in order.

Comments on the Introduction

Leonardo speaks first of how squares arise from the sums of consecutive odd numbers.

$$1 = 1^2.$$

$$1 + 3 = 2^2.$$

$$1 + 3 + 5 = 3^2.$$

$$1 + 3 + 5 + 7 = 4^2.$$

These are all instances of the general formula

$$1 + 3 + 5 + 7 + \cdots + (2N - 1) = N^2.$$

For example, the next formula giving the square of 5 is obtained by letting $N = 5$ in the general formula. When N is 5, then $2N - 1$ is the odd number 9.

$$1 + 3 + 5 + 7 + 9 = 5^2.$$

This result was known to Pythagoras by 500 B.C. [H2, p. 77].

Here is an easy and primitive argument for the truth of the formula.

$$\begin{aligned}
&1 + 3 + 5 + \cdots + (2N - 1) \\
&\quad = (1/2)[1 + 3 + 5 + \cdots + (2N - 3) + (2N - 1) \\
&\qquad + 1 + 3 + 5 + \cdots + (2N - 3) + (2N - 1)] \\
&\quad = (1/2)[1 + (2N - 1) + 3 + (2N - 3) + \cdots \\
&\qquad + (2N - 3) + 3 + (2N - 1) + 1] \\
&\quad = (1/2)[2N + 2N + 2N + \cdots + 2N] \\
&\quad = (1/2)[N(2N)] \\
&\quad = N^2.
\end{aligned}$$

This theorem is later restated as proposition 4 and Leonardo gives a proof at that place.

Proposition 1

[Find two square numbers which sum to a square number.]

Hence, to find two square numbers which sum to a square number, I shall take any odd square and I shall have it for one of the two said squares; the other I shall find in a sum of all odd numbers from unity up to the odd square itself. For example, I shall take 9 for one of the mentioned two squares, the other will be had in the sum of odd numbers which are smaller than 9, namely 1 and 3 and 5 and 7, which have sum 16, which is a square, which added to 9 will yield 25, which is a square number. And if we wish a geometric demonstration, any number of odd numbers from the unity in ascending order are adjoined, making the end be square; and let .*ab*. be 1, .*bc*. be 3, .*cd*. be 5, .*de*. be 7, .*ef*. be 9; and because .*ef*., 9, is a square and .*ae*. 16, is a square, created from the sum of the odd numbers .*ab*. and .*bc*. and .*cd*. and .*de*., the total number .*af*. is likewise square; and thus from the sum of the two squares .*ae*. and .*ef*. is made the square .*af*..

a b c d e f

Also, alternatively, I shall take some even square, and shall let half of it be also even, as 36 of which half is 18; and I shall take from it 1, and to it shall add 1, to yield 17 and 19, which are odd numbers and consecutive, with no odd number falling between them; their addition yields 36, which is square, and the addition of the remaining odd numbers from 1 up to 15 yields 64; the addition of the two squares yields 100, which is square, and is the sum of the odd numbers from 1 up to 19. As well, I shall take an odd square number, of which a third part is whole, as 81 of which a third is 27; and I shall take 27 itself with two odd numbers of which 27 is the mean, namely 25 and 29; and these three numbers sum to 81, which is square; and the sum of the others, which are from 1 up to 23, is 144, which has root 12; add then 144 to 81, from this comes a sum of odd numbers which are from 1 up to 29, namely 225, which is a square number and the root of it is 15. In a similar manner can be found four more consecutive odd numbers, the sum of which make a square number and the sum of the remaining smaller numbers down to unity yield also a square; and the two squares themselves always add to make a square number.

Comments on Proposition 1

Leonardo applies immediately the formula on the sum of consecutive odd numbers to get solutions to the Pythagorean problem: Find two square numbers which sum to a square number. Leonardo's first numerical example, which illustrates the principle, is

$$\begin{aligned}(1 + 3 + 5 + 7) + 9 &= (1 + 3 + 5 + 7 + 9).\\ 4^2 \quad + 3^2 &= \quad 5^2.\end{aligned}$$

With line segments representing successive odd numbers, Leonardo argues that

$$\begin{aligned}(ab + bc + cd + de) + ef &= (ab + bc + cd + de + ef).\\ ae \quad + ef &= \quad af.\end{aligned}$$

In the equations *ae*, *ef* and *af* are all square numbers.

The same argument in modern notation goes as follows. Let the square of $(2n - 1)$ represent any odd number squared. Noting that odd numbers squared make odd numbers and even numbers squared make even numbers, the odd numbers coming before the square of $(2n - 1)$ are

$$1, 3, 5, \ldots, (2n - 1)^2 - 2.$$

We then have this identity,

$$\begin{aligned}&[1 + 3 + 5 + \cdots + (2n - 1)^2 - 2]\\ &\qquad + (2n - 1)^2 = 1 + 3 + \cdots + (2n - 1)^2,\end{aligned}$$

which after we sum the odd numbers gives us

$$[2n^2 - 2n]^2 + (2n - 1)^2 = [2n^2 - 2n + 1]^2$$

We identified N of $2N - 1$ in the consecutive odd number sums using these calculations:

$$(2n - 1)^2 - 2 = 4n^2 - 4n - 1 = 2(2n^2 - 2n) - 1.$$

$$(2n - 1)^2 = 4n^2 - 4n + 2 - 1 = 2(2n^2 - 2n + 1) - 1.$$

In the example given by Leonardo, $n = 2$:

$$4^2 + 3^2 = 5^2.$$

If we substitute $n = 3$, we get

$$12^2 + 5^2 = 13^2.$$

The argument using an even square instead of an odd square to find a Pythagorean triple runs as follows: Let $(2n)^2$ be an even number squared. Half of it is $2n^2$. Two adjacent odd numbers are $2n^2 - 1$ and $2n^2 + 1$. Their sum is the even square $(2n)^2$. Since

$$2n^2 - 3 = 2n^2 - 2 - 1 = 2(n^2 - 1) - 1,$$

and

$$2n^2 + 1 = 2n^2 + 2 - 1 = 2(n^2 + 1) - 1.$$

we have

$$[1 + 3 + \cdots + (2n^2 - 3)] + [2n^2 - 1 + 2n^2 + 1]$$
$$= [1 + 3 + \cdots + (2n^2 + 1)].$$

$$(n^2 - 1)^2 + (2n)^2 = (n^2 + 1)^2.$$

If $n = 3$, then we have $8^2 + 6^2 = 10^2$.

Three consecutive odd numbers are used to build another solution. Let the odd square divisible by three be $9(2n + 1)^2$. Three consecutive odd numbers are

$$3(2n + 1)^2 - 2,\ 3(2n + 1)^2,\ 3(2n + 1)^2 + 2.$$

Since

$$3(2n + 1)^2 - 4 = 12n^2 + 12n + 3 - 4 = 2(6n^2 + 6n) - 1$$

and

$$3(2n + 1)^2 + 2 = 12n^2 + 12n + 5 = 2(6n^2 + 6n + 3) - 1$$

we have

$$[1 + 3 + 5 + \cdots + 3(2n + 1)^2 - 4]$$
$$+ [3(2n + 1)^2 - 2 + 3(2n + 1)^2 + 3(2n + 1)^2 + 2]$$
$$= [1 + 3 + 5 + \cdots + 3(2n + 1)^2 + 2].$$

$$(6n^2 + 6n)^2 + [3(2n + 1)^2] = (6n^2 + 6n + 3).$$

If $n = 1$, then we obtain

$$12^2 + 9^2 = 15^2.$$

Although Leonardo does not mention it, 9 is a factor of every term in the identity. Dividing through by 9 yields

$$(2n^2 + 2n)^2 + (2n + 1)^2 = (2n^2 + 2n + 1)^2,$$

which is identical with the earlier result. The formula obtained from using four odd numbers also will not give new information.

Proposition 2

[Any square number exceeds the square immediately before it by the sum of the roots.]

Similarly, I have found that any square exceeds the square immediately before it by the sum of the roots of these squares. For example, 121, of which the root is 11, exceeds 100, of which the root is 10, by the sum of 10 and 11, namely the sum of the roots themselves. This is why one square exceeds the second square before it by the quantity which is four times the root of the square which is between them, as 121, which exceeds 81 by four times 10; and thus can be found differences between squares by the distances between the roots themselves. And when consecutive roots added make a square number, then the square of the greater root is equal to the sum of two squares. Likewise, when four times another root is square, then the square of the following root is equal to the sum of two squares, one of which will be that created by the mentioned quadruple and the other is that which has root one less than the quadrupled root. Therefore, if 9 is quadrupled, then 36 is created. Thus 100, of which the root is 10, is equal to the sum 64, which has root 8, and 36, which was the quadruple of 9. And it is noted that out of the quadrupling of any number a square is obtained only if the number itself was a square because, as

Euclid has shown, when the ratio of a number to another number is the same as the ratio of squares, then as the square is made from multiplication, and because 4 is a square, the number that it multiplies should also be square in order to make a square. And thus in many ways we can find three square numbers so that one always is equal to the sum of the other two.

But it appears that every square exceeds its preceding square, as we said, by as much as the sum of the roots themselves, which will be evident if we place the roots on the segments $.ab.$ and $.bg.$. And since $.ab.$ and $.bg.$ are consecutive numbers, one will be bigger than the other by one. Let then $.bg.$ be bigger than $.ab.$ by one, and subtract the unity $.dg.$ from $.bg.$, and there will remain $.bd.$, equal to $.ba.$;

a *b* *d* *g*

and since $.bg.$ is a number divided into two parts, namely $.bd.$ and $.dg.$; $.dg.$ the product of $.bd.$ by itself added to the product of $.dg.$ by itself added to twice $.bd.$ times $.dg.$ will be equal to the product of $.bg.$ with itself. But the product of $.bd.$ with itself is equal to the product of $.ab.$ with itself. Therefore, the square of the number $.bg.$ exceeds that of the number $.ab.$ by the quantity which is the sum of $.gd.$ times itself and twice $.gd.$ times $.bd.$. But the product of $.dg.$ with itself is one, which equals and is the same as the unity $.dg.$; and twice $.dg.$ times $.bd.$ make twice $.bd.$, as $.dg.$ is 1; therefore, twice $.bd.$ is $.ad.$; therefore, the square of the number $.bg.$ exceeds the square made by the number $.ab.$ by a quantity which is the sum of the roots themselves, which are $.ab.$ and $.bg.$. This is what had to be demonstrated.

Alternatively, since the number $.bd.$ equals the number $.ba.$, the total $.ad.$ will be divided into two equal parts by the point $.b.$; and to $.ad.$ is added the unity $.dg.$; then the product of $.dg.$ by $.ag.$, added to the square of the root $.ab.$, will equal the square made by the root *.bg.*; this is why the square of the number *.bg.* exceeds the square of the number *.ab.* by that which is the product of *.dg.* times *.ag.*. But *.dg.* multiplied by *.ag.* makes the number *.ag.*, since *.dg.* is 1. Therefore, the square of *.bg.* exceeds the square of *.ab.* by the sum of the roots themselves, which sum is the number *.ag.*.

Similarly, it is demonstrated that any square exceeds any smaller square by the product of the difference of the roots by the sum of the roots. For example, let *.ag.* and *.gb.* be two roots of any two squares whatsoever, and let *.gb.* be bigger than *.ag.* by the number *.db.*. Because the product of *.ag.* with itself, plus the product of *.db.* with *.ab.*, equals the product of *.gb.* with itself, the square of *.gb.* exceeds the square of *.ag.* by as much as the root *.gb.* exceeds the root *.ag.* multiplied by the sum of *.gb.* and *.ag.*, namely, by the product of *.db.* and *.ab.*. This is what had to be demonstrated.

Comments on Proposition 2

The reference that Leonardo makes to Euclid is most likely to the *Elements*, Book VIII, proposition 24: If two numbers have to one another the ratio which a square number has to a square number, and the first be square, the second will also be a square [H3 vol. 2, p. 380].

Any square number exceeds the square immediately before it by the sum of the roots. The difference between consecutive squares is equal to the sum of the roots. In brief,

$$(n + 1)^2 - n^2 = 2n + 1 = (n + 1) + n;$$

$(n + 1)$ and n, of course, are the roots of $(n + 1)^2$ and n^2.

$$11^2 - 10^2 = 21 = 11 + 10.$$

If consecutive roots added make a square, then there is a Pythagorean triple; $(n + 1)^2 - n^2 = (n + 1) + n = y^2$, for some number y. Then $(n + 1)^2 = n^2 + y^2$. For example, if $n = 12$, then $(n + 1) + n = 25$, a square. This gives $13^2 = 12^2 + 5^2$.

When two roots differ by two, then their squares differ by four times the intermediate root.

$$(n + 2)^2 - n^2 = 4n + 4 = 4(n + 1).$$

$$11^2 - 9^2 = 121 - 81 = 40 = 4(10).$$

If roots differ by two and the intermediate root is a square, then there is a Pythagorean triple:

$$(n + 2)^2 - n^2 = 4(n + 1).$$

Since 4 and $4(n + 1)$ are both squares and 1 is also a square, it follows that $n + 1$ is a square. For example, let $n + 1 = 9$.

$$10^2 - 8^2 = 4(9).$$

$$10^2 = 8^2 + 6^2.$$

Here is Leonardo's proof using line segments but written with equations. Let ab be any number and bg the following number. Place d so that $dg = 1$ and $bd = ab$.

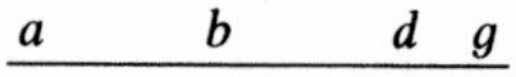

a b d g

$$\begin{aligned} bg^2 &= bd^2 + dg^2 + 2(bd)(dg) \\ &= ab^2 + (dg)(1) + 2(bd)(1) \\ &= ab^2 + dg + ad \\ &= ab^2 + ag \\ &= ab^2 + ab + bg. \end{aligned}$$

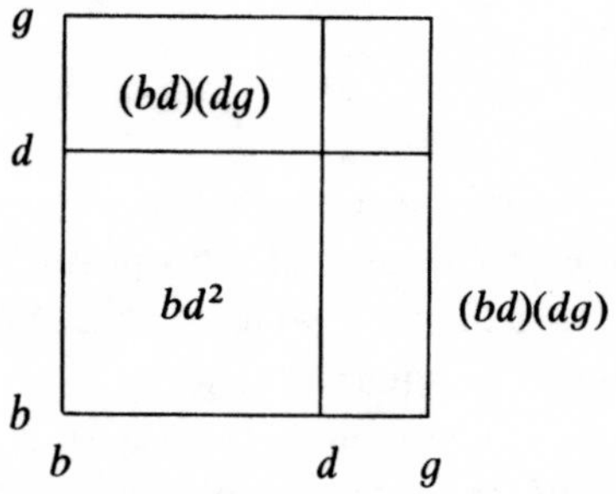

g
(bd)(dg)
d
bd^2
(bd)(dg)
b
b d g

Alternatively,

$$\begin{aligned} bg^2 &= bd^2 + (bd)(dg) + (bd)(dg) + dg^2 \\ &= ab^2 + (ab)(dg) + (bd)(dg) + dg^2 \\ &= ab^2 + (ab + bd + dg)dg \\ &= ab^2 + (ag)(dg) \\ &= ab^2 + ag \\ &= ab^2 + ab + bg. \end{aligned}$$

Finally, we look at the result that the difference between any two squares whatsoever is equal to the product of the sum of the roots and the difference of the roots. In modern notation, simply

$$m^2 - n^2 = (m + n)(m - n).$$

Here is Leonardo's argument with line segments. Let ab and gb be any two roots and construct $ag = gd$. Therefore, $gb = bd + db$.

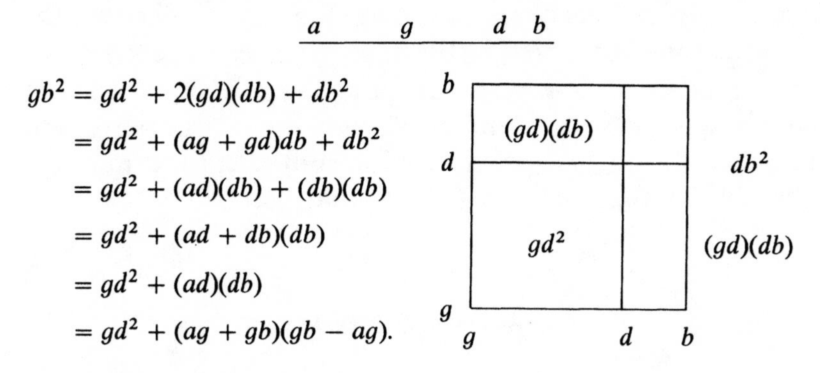

$$\begin{aligned} gb^2 &= gd^2 + 2(gd)(db) + db^2 \\ &= gd^2 + (ag + gd)db + db^2 \\ &= gd^2 + (ad)(db) + (db)(db) \\ &= gd^2 + (ad + db)(db) \\ &= gd^2 + (ad)(db) \\ &= gd^2 + (ag + gb)(gb - ag). \end{aligned}$$

These arguments in geometrical algebra are found in Book II of Euclid's *Elements* [H3].

Proposition 3

[There is another way of finding two squares which make a square number with their sum.]

There is indeed another way of finding two squares which make a square number with their sum, and it is found in Book X of Euclid. Adjoin two square numbers both even or both odd, $.ab.$ and $.bg.$; then the sum $.ag.$ will be even. Let $.ab.$ be bigger than $.bg.$, and $.ag.$ is divided in two equal parts by $.d.$. The number $.ad.$ is then a whole number because it is half of the number *.ag.*. And subtract *.ad.* from the number *.ab.*; there will remain the whole number *.db.*.

a *d* *b* *g*

And because the number *.ag.* is divided into equal parts by *.d.*, and into unequal parts by *.b.*, the product of *.ab.* and *.bg.*, plus the square of the number *.db.*, will equal the square of the number *.dg.*; but that which is made from *.ab.* times *.bg.* is a square, as *.ab.* and *.bg.* are squares; that which is made by the number *.db.* times *.db.* is a square, and thus are found two squares with sum a square number, namely the number *.dg.*. This is what had to be done.

Comments on Proposition 3

The Pythagorean triples given by this proposition have the form

$$[(m^2 + n^2)/2]^2 = [(m^2 - n^2)/2]^2 + m^2n^2.$$

That this formula is correct can be easily verified by multiplication. Proposition 5 of Book II of Euclid's *Elements* is a direct reference for the equation used by Leonardo [H3, vol. 1, p. 382]. Proposition 5 is essentially this equation

$$[(x + y)/2]^2 = [(x - y)/2]^2 + xy.$$

This equation was known even to the ancient Babylonian mathematicians and was used to solve quadratic equations. The reference given by Leonardo is to Book X of Euclid's *Elements.* This is most probably to lemma 1 of proposition 29 of Book X, which includes the same information.

But here is Leonardo's argument. Let *ag* and *bg* be given square numbers so that they are either both odd or both even.

a *d* *b* *g*

Choose d so that $ad = dg$, that is d bisects the even number ag. ad, which equals $(\frac{1}{2})ag$, is a whole number. $ab = ad + db$.

$$\begin{aligned} dg^2 &= db^2 + 2(db)(bg) + bg^2 \\ &= db^2 + [2(db) + bg]bg \\ &= db^2 + (db + db + bg)bg \\ &= db^2 + (db + dg)bg \\ &= db^2 + (db + ad)bg \\ &= db^2 + (ab)(bg). \end{aligned}$$

Because ab and bg are given squares, $(ab)(bg)$ is a square number and a Pythagorean triple is obtained. The more conventional formula is obtained by letting $ab = m^2$ and $bg = n^2$.

$$dg^2 = [(1/2)(ag)]^2 = [(1/2)(ab + bg)]^2 = [(1/2)(m^2 + n^2)].$$

$$\begin{aligned} db^2 &= (ab - ad)^2 = [m^2 - (m^2 + n^2)/2]^2 \\ &= [(m^2 - n^2)/2]^2. \end{aligned}$$

$$(ab)(bg) = m^2n^2.$$

That ab and bg have the same parity ensures that both $(\frac{1}{2})(ag)$ and $(\frac{1}{2})(db)$ are whole numbers.

Proposition 4

I wish to demonstrate how a sequence of squares is produced from the ordered sums of odd numbers which run from 1 to infinity.

Adjoin, beginning with the unity *.ab.*, any number of consecutive numbers, *.bg.*, *.gd.*, *.de.*, *.ez.*, *.zi.*; join together *.bg.* with *.ab.* to make the number *.t.*; similarly join together each number with its antecedent and with its successor; and let *.k.* be the sum of the numbers *.bg.* and

.gd.; even join the numbers *.gd.* and *.de.* to make the number *.l.*; also the numbers *.de.* and *.ez.* to make the number *.m.*; and *.n.*, namely, the sum of *.ez.* and *.zi.*. I say first that *.t.*, *.k.*, *.l.*, *.m.*, *.n.* are consecutive odd numbers beginning with unity. Certainly the number *.zi.* is either even or odd; if the number *.zi.* is even, then the number *.ez.* is odd; and if the number *.zi.* is odd, then the number *.ez.* is even; for these numbers certainly are consecutive. Therefore, the sum of the numbers *.ez.* and *.zi.*, namely *.n.*, is odd. Similarly, we shall show the sum of the numbers *.de.* and *.ez.*, namely *.m.*, is odd. By the same method, the numbers *.l.*, *.k.*, *.t.* will be shown to be odd; I say, in fact, that *.t.*, *.k.*, *.l.*, *.m.*, *.n.* are consecutive odd numbers. In fact, the number *.n.* is made from the addition of *.ez.* with *.zi.*; and from the addition of *.de.* with *.ez.* is made the number *.m.*. As much, therefore, as the number *.zi.* exceeds the number *.de.*, the number *.n.* exceeds the number *.m.*. In truth, the number *.zi.* exceeds the number *.ez.* by one, the same by which the number *.ez.* exceeds the number *.de.*.Therefore, the number *.zi.* exceeds the number *.de.* by two. Therefore, the number *.n.* exceeds the number *.m.* similarly by two; in the same manner, it will be found that the number *.m.* exceeds the number *.l.*, and the number *.l.* the number *.k.*, and the number *.k.* the number *.t.*, and the number *.t.* the unity *.ab.*. Therefore, the unity and *.t.*, *.k.*, *.l.*, *.m.*, *.n.*, are, as we predicted, consecutive odd numbers.

And, as is shown above, the square which is made by the number *.zi.* exceeds the square which is made by the number *.ez.* by a number which is the sum of *.zi.* and *.ez.*; this is the number *.n.*. Similarly, it is shown that the square made by the number *.ez.* exceeds the square made by the number *.de.* by the sum of the numbers *.de.* and *.ez.*; this is number *.m.*. And the square made by the number *.de.* exceeds the square made by the number *.gd.*, the number *.l.*. And the square made by the number *.gd.* exceeds the square made by the number *.bg.* by *.k.*. And the square of the number *.bg.* exceeds the square of the unity by the number *.t.*; *.t.* is certainly 3, and *.bg.* is 2. Therefore, if to the square of unity, that is 1, is added the number *.t.*, by which the square of *.bg.* exceeds the square of unity, the square of the number *.bg.* is obtained; if to that is added the number *.k.*, the square of the number *.gd.* will result; if to this

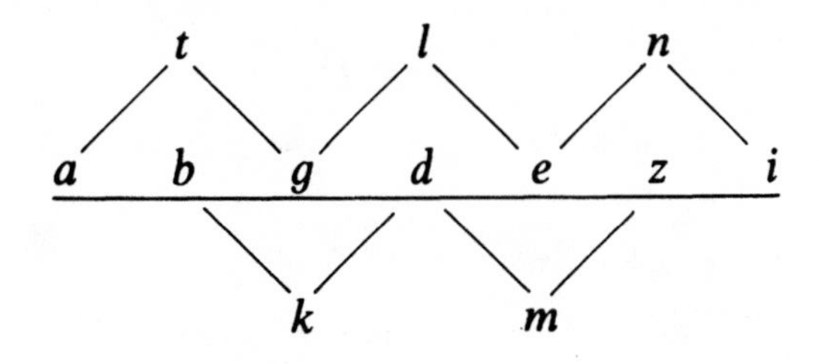

square is added the number *.l.*, the square of the number *.de.* will result; if to this square is added the number *.m.*, the square of the number *.ez.* will result; if to this again is added the number *.n.*, by which the square of the number *.zi.* exceeds the square of the number *.ez.*, clearly will result the square of the number *.zi.*. Certainly the numbers *.ab.*, *.bg.*, *.gd.*, *.de.*, *.ez.*, *.zi.* are consecutive and their squares arise from the sum of consecutive odd numbers *.ab.*, *.t.*, *.k.*, *.l.*, *.m.*, *.n.* as was to be shown.

Comments on Proposition 4

In this proposition Leonardo gives a proof of the formula he set out in the introduction,

$$1 + 3 + 5 + \cdots + (2s - 1) = s^2,$$

based upon the difference of consecutive squares equaling the sum of the roots. The same argument in modern notation is given in the right column parallel to Leonardo's in the left column.

Consecutive whole numbers *ab*, *bg*, *gd*, *de*, *ez*, *zi* are given. We find the sum $ab + bg + gd + de + ez + zi$ or, in modern notation, $1 + 3 + 5 + \cdots + (2s + 1)$.

$$\begin{array}{ll} ab + bg = t. & 1 + 2 = 3. \\ bg + gd = k. & 2 + 3 = 5. \\ gd + de = 1. & 3 + 4 = 7. \\ de + ez = m. & \ldots \\ ez + zi = n. & (s - 1) + s = 2s - 1, \end{array}$$

where 1, t, k, l, m, n and 1, 3, 5,..., $(2s - 1)$ are consecutive odd numbers.

$$\begin{array}{ll} zi^2 - ez^2 = zi + ez = n. & s^2 - (s - 1)^2 \qquad = s - 1. \\ ez^2 - de^2 = ez + de = m. & (s - 1)^2 - (s - 2)^2 = 2s - 3. \\ de^2 - gd^2 = de + gd = l. & (s - 2)^2 - (s - 3)^2 = 2s - 5. \\ gd^2 - bg^2 = gd + bg = k. & \ldots \\ bg^2 - 1^2 = bg + 1 = t. & 2^2 - 1^2 \qquad = 3. \\ 1^2 = \quad 1. & 1^2 \qquad = 1. \end{array}$$

The result is obtained by summing the colums.

$$zi^2 = 1 + t + k + l + m + n. \qquad s^2 = 1 + 3 + 5 + \cdots + (2s - 1).$$

In the Latin manuscript, *.A.* is sometimes used for the unity as well as *.ab.*. In this English translation, *.ab.* is used uniformly.

Proposition 5

Find two numbers so that the sum of their squares makes a square formed by the sum of the squares of two other given numbers.

Let two numbers *.a.* and *.b.* be given so that the sum of their squares makes a square number *.g.*; one must find two other numbers so that the sum of their squares equals the square number *.g.*.

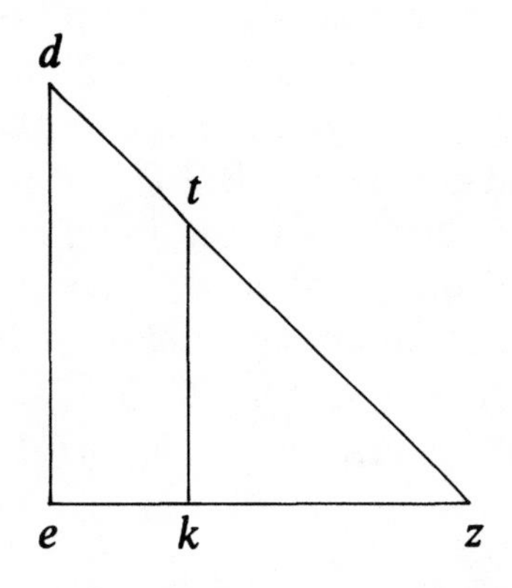

Let any other two numbers be found so that the sum of their squares is a square number. These two numbers are represented with segments *.de.* and *.ez.*, and are put together at a right angle, which is, namely, the angle *.dez.*. Also, the segment *.dz.* is located against the sides *.de.* and *.ez.*. The square made by the segment *.dz.* is equal to the number *.g.* or not. First, if equal, then the two other numbers for which the sum of their squares equals *.g.* are found, one of these numbers is equal to the segment *.de.* and the other to the segment *.ez.*. If not, if the square made by the segment *.dz.*, that is the number *.dz.*, is not equal to the number *.g.*, it will be either bigger or smaller than *.g.*. First, if bigger, the square made by the number *.dz.* is bigger than the number *.g.*, then the number *.dz.* will be bigger than the square root of *.g.*; therefore, the root of the number *.g.* is taken equal to the number *.i.*, and is placed upon the length *.dz.*, and is denoted by *.tz.*. And from the point *.t.* draw *.tk.* perpendicular to *.ez.*; *.tk.* is therefore parallel to *.de.*. Because the triangle *.tkz.* is similar to the triangle *.dez.*, *.zd.* is therefore to *.zt.* as *.de.* is to *.tk.*. But the ratio of *.zd.* to *.zt.* is known; both lengths are indeed known. Because of this, the ratio of .*de*. to .*tk*. will be similarly known. And .*de*. is known. Therefore, the segment .*tk*. will be known. Similarly, it is shown that the segment .*zk*. is known with the ratio of it to .*ze*. as .*zt*. to .*zd*.; known therefore are .*tk*. and .*kz*., which have the sum of their squares equal to the square made by segment .*tz*.. But the square of the number .*tz*. is equal to the square of the number .*i*., and .*i*. is indeed the square root of the number .*g*.. Therefore, the square of .*tz*. is equal to the number .*g*.; two numbers .*tk*. and .*kz*. are indeed found with the sum of their squares equal to the square number .*g*.. Alternatively, let .*dz*. be smaller than .*i*. and extend the line

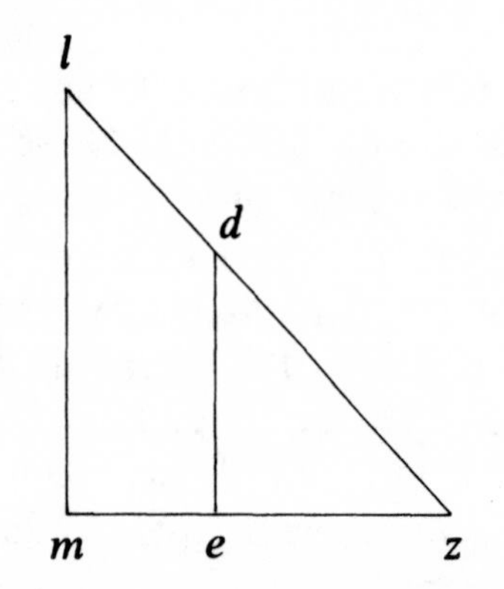

.*zd*. up to .*l*. and put .*zl*. equal to the number .*i*.. Similarly, .*ze*. is extended and .*lm*. is connected so that .*lm*. is parallel to .*de*.; therefore, the triangle .*dez*. is similar to the triangle .*lmz*., and the ratio .*zd*. to .*zl*. is known. Therefore, both numbers .*zm*. and .*ml*. will be known. The two numbers .*lm*. and .*mz*. are found and the sum of their squares equals the number .*g*., with .*lz*. equal to the root. This is what had to be done.

But in order to have this in numbers. let .*a*. be 5 and .*b*. be 12. Then .*g*., which is the sum of the squares of the numbers .*a*. and .*b*., is 169, and its root, namely .*i*., is 13. Join the two segments .*de*. and .*ez*. at a right angle .*dez*.;

and let the segment *.de.* be 15 and the segment *.ez.* be 8; consequently, *.dz.* will be 17. Put along the line *.dz.* the segment *.zt.* equal to *.i.*; *.zt.* is then 13, and *.tk.* is drawn parallel to *.de.*; therefore, *.zd.* is to *.zt.* as *.de.* is to *.tk.*. Multiply therefore *.zt.* by *.de.*, that is 13 by 15, and divide the product by *.dz.*, that is by 17, producing the number $11\frac{8}{17}$ for *.tk.*. Similarly, if *.zt.* shall be multiplied by *.ze.*, and divided by *.zd.*, $6\frac{2}{17}$ is produced for *.kz.*; and thus are found two numbers, namely *.tk.* and *.kz.*, which have the sum of their squares equal to the number *.g.*, that is the square of *.zt.*. Likewise, it is shown that if the number *.dz.* is smaller than *.i.* as in the other figure in which we shall put 4 for *.de.* and 3 for *.ez.*. Therefore, *.dz.* is 5; and *.zd.* is extended to *.l.*, and *.zl.* equals *.i.*, namely 13; and as *.zd.* is to *.zl.* so is *.de.* to *.lm.*. Therefore, multiply *.zl.* by *.de.* and divide by *.zd.*, producing $10\frac{2}{5}$ for *.lm.*. Similarly, multiply *.zl.* by *.ze.*, divide by *.zd.*, namely 39 by 5, producing $7\frac{4}{5}$ for *.mz.*. And thus are found another two numbers with the sum of their squares the same 169; they are $10\frac{2}{5}$ and $7\frac{4}{5}$. And thus is shown how it can be done in an infinite number of ways.

Comments of Proposition 5

Find two numbers so that the sum of their squares makes a square number which is the sum of the two given square numbers. In arguing with line segments, Leonardo gives his argument in three cases depending upon whether one segment exceeds, falls short of, or is equal to a given segment. One algebraic argument in modern notation will suffice for all cases.

Let a and b be the given numbers and let the sum of their squares be g or the square of i. Let de and ez be any other two numbers and let the sum of their squares be the square of dz.

$$a^2 + b^2 = g = i^2.$$

$$de^2 + ez^2 = dz^2.$$

If $dz^2 = g = i^2$, then de and ez are the solution. If $dz^2 > g$, then set $g = tz^2$. $dz/tz = de/tk$.

Leonardo simply uses proportional parts of de and ez to sum to the square g. These are shown on the triangle diagram.

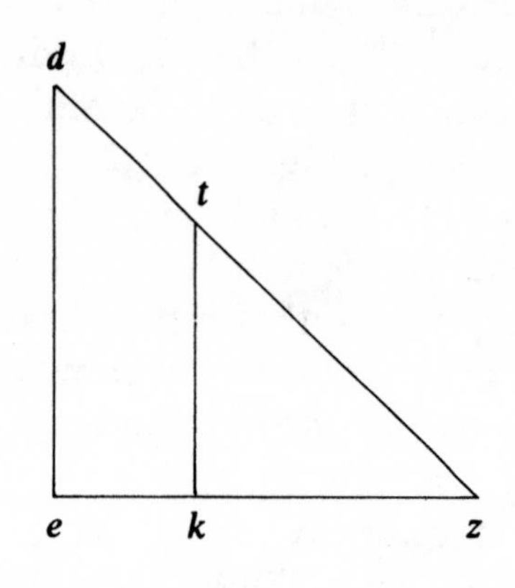

Then

$$tk^2 + kz^2 = tz^2.$$

$$[(de/dz)tz]^2 + [(ez/dz)tz]^2 = g.$$

The case $dz < g$ is similar.

The result in modern notation is as follows:

$$a^2 + b^2 = i^2 \qquad \text{Let } m^2 + n^2 = p^2.$$

$$(m/p)^2 + (n/p)^2 = 1.$$

$$[(mi)/p]^2 + [(ni)/p]^2 = i^2 = g.$$

Here is an example to find the square of 13 in proportional parts of 8 and 15.

$$5^2 + 12^2 = 13^2. \qquad 8^2 + 15^2 = 17^2.$$

$$[8(13/17)]^2 + [15(13/17)]^2 = 8^2 13^2/17^2 + 15^2 13^2/17^2 = 13^2.$$

$$(104/17)^2 + (195/17)^2 = 13^2.$$

$$(6\tfrac{2}{17})^2 + (11\tfrac{8}{17})^2 = 13^2.$$

Here is another example.

$$5^2 + 12^2 = 13^2. \qquad 3^2 + 4^2 = 5^2.$$

$$[4(13/5)]^2 + [3(13/5)]^2 = 13^2.$$

$$(52/5)^2 + (29/5)^2 = 13^2.$$

$$(10\tfrac{2}{5})^2 + (7\tfrac{4}{5})^2 = 13^2.$$

Proposition 6

If four numbers, not in geometric proportion, are given, and if the first is smaller than the second and the third smaller than the fourth, and if the sum of the squares of the first and the second is multiplied by the sum of the squares of the third and fourth, and neither of these sums make a square, a number is obtained which is equal to the sum of two squares in two ways; and if one of these sums is a square, then the number obtained is a sum of squares in three ways; and if both sums are squares, then the number obtained is a sum of squares in four ways; and this is understood to be without fractions.

Let four nonproportional numbers *.a.*, *.b.*, *.g.*, *.d.* be given, and let *.a.* be smaller than *.b.* and *.g.* smaller than *.d.*; and let the sum of the squares of *.a.* and *.b.* be the number *.e.*, and the sum of the squares of *.g.* and *.d.* be *.z.*, and *.e.* be multiplied by *.z.* producing the number *.cf.*; and let neither of the numbers *.e.* and *.z.* be a square. I say that this number *.cf.* is equal to the sum of two squares, even in two ways. First *.a.* is multiplied by *.g.*, and the number

t k p l

.tk. is obtained; and out of *.b.* times *.d.* is obtained *.kl.*; and out of *.a.* times *.d.* is obtained *.mn.*; and out of *.b.* times *.g.* is obtained *.no.*. And since the numbers *.a.*, *.b.*, *.g.*, *.d.*, are not proportional, and *.a.* is smaller than *.b.* and *.g.* is smaller than *.d.*, the aforementioned products are necessarily unequal, and *.tk.* is smaller than *.kl.*; let *.kp.* be on *.kl.* so that *.kp.* equals *.tk.*. Similarly for the numbers *.mn.* and *.no.*, let *.no.* be bigger than *.mn.*, and take *.nq.* equal to *.mn.*. I say that

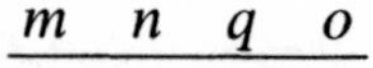

m n q o

the number *.cf.* is equal to the sum of the squares made by the numbers *.tl.* and *.qo.*, and by the numbers *.mo.* and *.pl.*, because *.e.* times *.z.* produces *.cf.* and *.e.* is the sum of the two squares made by *.a.* and *.b.*. Therefore, *.cf.* results from the multiplication of the square of the number *.a.* by *.z.*, added to the square of the number *.b.* multiplied by *.z.*. Let then *.ci.*

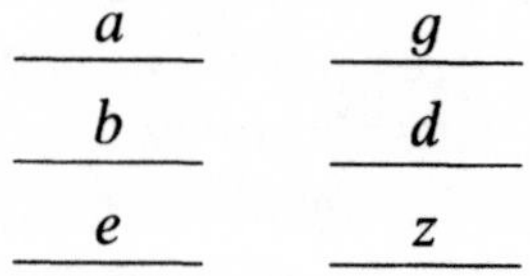

a *g*

b *d*

e *z*

be that which results from the multiplication of *.a.* with itself times *.z.*, namely the square of *.a.* times *.z.*; there will remain therefore *.if.* for the number which results from multiplying the

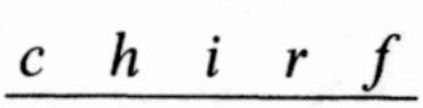

c h i r f

square of *.b.* with *.z.*. But *.z.* is the sum of squares of the numbers *.g.* and *.d.*. Consequently, the product of the square of *.a.* with *.z.* is equal to the sum of two products, namely the square of *.a.* times the square of the number *.g.*, and the square of *.a.* times the square of the number *.d.*. Let, therefore, *.ch.* be that which results from the multiplication of the square of *.a.* with the square of the number *.g.*;

therefore, *.hi.* will be that which result s from the multiplication of the square of the number *.a.* with the square of the number *.d.*. Again, let *.ir.* be the product of the number *.b.* and itself with the square of the number *.g.*; there will remain, therefore, *.rf.*, which is the product of the square *.b.* with the square of the number *.d.*. Therefore, the total number *.cf.* is divided into four numbers which are *.ch.*, *.hi.*, *.ir.*, *.rf.*; and each one of them is a square made from the multiplication of a square number with a square number, having roots I shall show to be *.tk.*, *.kl.*, *.mn.*, *.no.*; the first, the number *.ch.*, I shall show to be equal to the square of the number *.tk.*; certainly *.ch.* is made from the multiplication of the square of the number *.a.* with the square of the number *.g.*. But *.tk.* is the product of *.a.* and *.g.*; consequently, the square of *.tk.* is equal to the square of the product of *.a.* and *.g.*. Similarly, it is shown that the square of the product of *.a.* and *.d.* is equal to the square of the number *.mn.*, which is equal to the number *.hi.*; and the square of the number *.kl.* is the number *.ir.*; and the square of the number *.no.* is equal to the number *.rf.*. It remains, indeed, to show that the sum of the two squares of the numbers *.tl.* and *.qo.*, as well as the sum of the squares of the numbers *.mo.* and *.pl.*, are equal to the sum of the four squares of the numbers *.tk.*, *.kl.*, *.mn.*, *.no.*. I shall show first the equality for the sum of the squares of the numbers *.tl.* and *.qo.*. The square of the number *.tl.* is in fact equal to the sum of two of the mentioned four squares, those made by the numbers *.tk.* and *.kl.* plus twice the product of *.tk.* and *.lk.*. Consequently, it remains to demonstrate that twice the product of *.tk.* and *.kl.* plus the square of the number *.qo.* equals the sum of the two remaining squares, namely those made by *.mn.* and *.no.*. First I shall show that *.tk.* times *.kl.* is equal to *.mn.* times *.no.*.

$\overline{\quad a \quad}$

In fact, .*tk*. results from the multiplication of .*a*. and .*g*., and .*kl*. results from .*b*. times .*d*.. Therefore, .*tk*. times .*kl*. results from .*a*. times .*g*. multiplied by .*b*. times .*d*.. Similarly, the product of .*mn*. and .*no*. arises from .*a*. times .*d*. multiplied by .*b*. times .*g*.. Therefore, .*mn*. times .*no*. is .*tk*. times .*kl*.. Therefore, it must be shown that twice .*mn*. times .*no*. plus the square of the number .*qo*. is equal to the sum of the squares of the numbers .*mn*. and .*no*.. Certainly .*nq*. equals .*mn*.; therefore, the square of the number .*mn*. is equal to the product of .*mn*. and .*nq*.; for .*mn*. times .*no*. exceeds .*mn*. times .*nq*. by that which is .*mn*. times .*qo*.. Therefore, the product of .*mn*. by .*no*. exceeds the square of the number .*mn*. by as much as .*qo*. times .*mn*., that is .*qo*. times .*qn*.. And because the number .*mn*. equals the number .*qn*., to one and the other is added .*qo*.. Therefore, the total .*no*. will equal the sum of the numbers .*mn*. and .*qo*.. Therefore, the square of the number .*no*. is equal to the sum of the two products .*on*. times .*nm*. and .*on*. times .*oq*.. Therefore, the square of the number .*no*. exceeds the product of .*on*. and .*nm*. by .*qo*. times .*on*.. But the product of .*mn*. and .*no*. exceeds the square of the number .*mn*. by .*nq*. times .*qo*.. But the square of the number .*no*. exceeds the product of .*mn*. and .*no*. by that which is .*no*. times .*oq*.. But the product of .*no*. and .*qo*. exceeds the product of .*oq*. and .*qn*. by that which is the number .*qo*. times itself. Therefore, the sum of the squares of the numbers .*mn*. and .*no*. exceeds twice .*mn*. times .*no*., that is .*tk*. times .*kl*., by the square of the number .*qo*.. But twice the product of .*tk*. and .*kl*. plus the square of the number .*qo*. is equal to the sum of the two squares of the numbers .*mn*. and .*no*.. Therefore, the sum of the squares of the numbers .*tl*. and .*qo*. is equal to the sum of the squares of the numbers .*tl*., .*kl*., .*mn*., .*no*., that is the number .*cf*.. This is what had to be shown.

From this is shown, in fact, that when two unequal numbers are given, twice the product of one with the other plus the square of the amount by which the bigger number exceeds the smaller number is equal to the sum of the squares of the same numbers. Therefore, twice the product of *.tk.* and *.kl.*, that is twice *.mn.* times *.no.*, added to the square of the number *.pl.* is equal to the sum of the squares of the numbers *.tk.* and *.kl.*. Therefore, if the two squares of the sumbers *.mn.* and *.no.* and twice the product *.mn.* times *.no.* and the three squares of the numbers *.pl.*, *.mn.*, *.no.*, are added, the sum will be equal to the sum of the four squares of the numbers *.tk.*, *.kl.*, *.mn.*, *.no.*, that is the number *.cf.*. But twice the product *.mn.* times *.no.* plus the sum of the squares of *.mn.* and *.no.* are equal to the square of the number *.mo.*. Therefore, the sum of the two squares of the numbers *.mo.* and *.pl.* is equal to the number *.cf.*, as had to be shown.

But let one of the numbers *.e.* and *.z.* be a square, and let it first be *.e.*. I say that it is possible to find two other numbers so that the sum of their squares is equal to *.cf.*, one of which results from the multiplication of *.e.* with the square of the number *.g.*, and the other results from the multiplication of *.e.* with the square of *.d.*. Because *.e.* is a square number, if it is multiplied by a square number the product will be square. Therefore, the squares of the numbers *.g.* and *.d.*, multiplied by the square *.e.*, will be squares. But the sum of the squares of the numbers *.g.* and *.d.* is *.z.*; and *.e.* times *.z.* produces *.cf.*, which had to be shown.

Similarly, if the numbers *.e.* and *.z.* are squares, there will be two other square numbers which add to make the number *.cf.*; and these result from multiplying *.z.* times the sum of the squares of the numbers *.a.* and *.b.* and from multiplying *.e.* times the sum of the squares of the numbers

.*g*. and .*d*.. And, as I said, if one of the numbers .*e*. and .*d*. is square, the number .*cf*. is equal to the sum of two different squares in three ways, and if both are square, is equal to the sum of two different squares in four ways.

Comments on Proposition 6

Let a, b, g, d be given so that $a < b$, $g < d$, $a/b \neq g/d$. Then the product

$$(a^2 + b^2)(g^2 + d^2)$$

is equal to the sum of two squares in two or three or four ways. Leonardo establishes two identities to prove his results. These equations are today called the Lagrange identities; obviously Leonardo has a prior claim to them. They were, however, used implicitly by Diophantus and mentioned by Arabic sources. In Book III of the *Arithmetica*, problem 19 [H1, p. 166], for example, Diophantus writes

$$65 = (13)(5) = (3^2 + 2^2)(2^2 + 1^2) = 8^2 + 1^2 = 7^2 + 4^2,$$

an application of the formula. The formula itself is explicitly stated by al-Khazin [A, p. 152] in approximately 950 A.D., and he actually discusses its use by Diophantus in problem 19 of Book III. Of course, it is not possible to know exactly what Arabic sources were available to Leonardo. See Mr. Kurt Vogel's article on Leonardo [Vo, p. 611] for a discussion on Leonardo's sources. Leonardo uses the identity to establish the theorem, his result on representing numbers as sums of squares. An informative history of the problem of representing numbers as sums of squares through the times of Fermat and Euler is given by Mr. André Weil in [We].

$$(a^2 + b^2)(g^2 + d^2) = [(a)(g) + (b)(d)]^2 + [(b)(g) - (a)(d)]^2 \quad (*1)$$

$$(a^2 + b^2)(g^2 + d^2) = [(a)(d) + (b)(g)]^2 + [(b)(d) - (a)(g)]^2 \quad (*2)$$

The geometrical algebra arguments with line segments put by Leonardo to get these results are involved. Let us first see how the conclusion of the theorem follows from the identities. Let e and z have the following values:

$$a^2 + b^2 = e. \qquad g^2 + d^2 = z.$$

The product of e and z is denoted by the segment cf. In case neither e nor z is itself a square, then cf is a sum of squares in two ways, as given by equations (*1) and (*2). If e itself is a square, then

$$(a^2 + b^2)(g^2 + d^2) = e(g^2 + d^2) = (e)g^2 + (e)d^2$$

gives a third way of writing cf as a sum of squares. If z is a square, then

$$(a^2 + b^2)(g^2 + d^2) = (a^2 + b^2)z = a^2(z) + b^2(z)$$

gives another way of writing cf as a sum of squares. Thus, cf is the sum of squares in two, three or four ways.

Here is a numerical example. Let $a = 6$, $b = 9$, $g = 3$, and $d = 4$.

$$(6^2 + 9^2)(3^2 + 4^2) = [(6)(3) + (9)(4)]^2 + [(9)(3) - (6)(4)]^2 \quad (*1)$$
$$= 54^2 + 3^2.$$

$$(6^2 + 9^2)(3^2 + 4^2) = [(6)(4) + (9)(3)]^2 + [(9)(4) - (6)(3)]^2 \quad (*2)$$
$$= 51^2 + 18^2.$$

$$(6^2 + 9^2)(3^2 + 4^2) = (6^2 + 9^2)5^2 = 6^2 5^2 + 9^2 5^2 = 30^2 + 45^2.$$

We will now write down the essential steps and equations in Leonardo's Euclidean geometrical algebra argument for equations (*1) and (*2). The segment cf is the product $(e)(z)$. $(a)(g) = tk$. $(b)(d) = kl$. $(a)(d) = mn$. $(b)(g) = no$.

	$(a)(d) \neq (b)(g)$.	$mn \neq no$.
$a < b$.	$(a)(g) < (b)(g)$.	$tk < no$.
	$(a)(d) < (b)(d)$.	$mn < kl$.
$g < d$.	$(a)(g) < (a)(d)$.	$tk < mn$.
	$(b)(g) < (b)(d)$.	$no < kl$.
$a < b$ and $g < d$.	$(a)(g) < (b)(d)$.	$tk < kl$.

$t \quad k \quad p \quad l$

Let $kp = tk$. $pl + tk = kl$.

$m \quad n \quad q \quad o$

Let $nq = mn$. $mn + qo = no$. $qo = no - mn$.

Here follows a statement of the two identities which must be proven.

$$cf = tl^2 + qo^2 = (tk + kl)^2 + (no - mn)^2$$
$$= [(a)(g) + (b)(d)]^2 + [(b)(g) - (a)(d)]^2. \qquad (*1)$$

$$cf = mo^2 + pl^2 = (mn + no)^2 + (kl - tk)^2$$
$$= [(a)(d) + (b)(g)]^2 + [(b)(d) - (a)(g)]^2. \qquad (*2)$$

$$cf = (a^2 + b^2)(g^2 + d^2).$$

$$cf = a^2z^2 + b^2z^2 = ci + if. \qquad z = g^2 + d^2.$$
$$a^2z^2 = a^2(g^2 + d^2) = a^2g^2 + a^2d^2 = ch + hi.$$
$$b^2z^2 = b^2(g^2 + d^2) = b^2g^2 + b^2d^2 = rf + ir.$$

$$cf = ch + hi + ir + rf = a^2g^2 + a^2d^2 + b^2d^2 + b^2g^2.$$
$$ch = a^2g^2 = [(a)(g)]^2 = tk^2.$$
$$hi = a^2d^2 = [(a)(d)]^2 = mn^2$$
$$ir = b^2d^2 = [(b)(d)]^2 = kl^2.$$
$$rf = b^2g^2 = [(b)(g)]^2 = no^2.$$

$$(tl)^2 + (qo)^2 = (tk)^2 + (mn)^2 + (kl)^2 + (no)^2.$$
$$(tl)^2 = (tk + kl)^2 = (tk)^2 + (kl)^2 + 2(tk)(kl).$$
$$(tk)(kl) = (a)(g)(b)(d) = (a)(d)(b)(g) = (mn)(no).$$
$$(tl)^2 = (tk + kl)^2 = (tk)^2 + (kl)^2 + 2(mn)(no).$$
$$nq = mn.$$
$$(tl)^2 + (qo)^2 = (tk)^2 + (kl)^2 + 2(mn)(no) + (qo)^2$$
$$= (tk)^2 + (kl)^2 + 2(mn)(nq + qo) + (qo)^2$$
$$= (tk)^2 + (kl)^2 + 2(mn)(nq) + 2(mn)(qo) + (qo)^2$$
$$= (tk)^2 + (kl)^2 + 2(mn)^2 + 2(nq)(qo) + (qo)^2$$
$$= (tk)^2 + (kl)^2 + (mn)^2 + (nq)^2 + 2(nq)(qo) + (qo)^2$$
$$= (tk)^2 + (kl)^2 + (mn)^2 + (nq + qo)^2$$
$$= (tk)^2 + (kl)^2 + (mn)^2 + (no)^2.$$

And similarly,

$$cf = (mo)^2 + (pl)^2 = (mn + no)^2 + (tk - kl)^2$$
$$= [(a)(d) + (b)(g)]^2 + [(a)(g) - (b)(d)]^2.$$

Proposition 7

Find in another way a square number which is equal to the sum of two square numbers.

Take four proportional numbers $.a.$, $.b.$, $.g.$, $.d.$ so that $.a.$ is to $.b.$ as $.g.$ is to $.d.$; and let $.e.$ be the sum of the squares of $.a.$ and $.b.$,

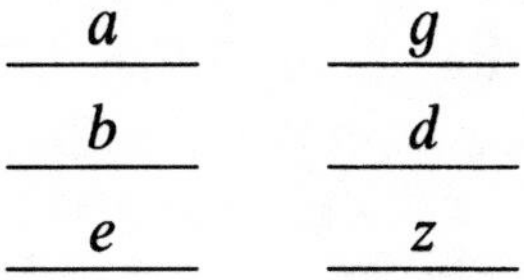

a g

b d

e z

and $.z.$ likewise with $.g.$ and $.d.$. And $.e.$ is multiplied by $.z.$, resulting in $.cf.$.

c f

Hence I say that $.cf.$ is a square and is equal to the sum of two squares, which is proved thus. From the product of $.a.$ times $.g.$ results, in fact,

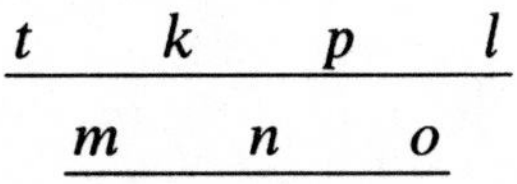

t k p l

m n o

$.tk.$, and from $.b.$ times $.d.$ results $.kl.$, and from $.a.$ times $.d.$ results $.mn.$, and from b times $.g.$ results $.no.$. I say first that the number *.mn.* is equal to the number *.no.*; for the numbers *.a.*, *.b.*, *.g.*, *.d.* are proportional in the ratio of *.a.* to *.b.*. Therefore, the product of *.a.* times *.d.* is equal to the product of *.b.* times *.g.*; that is, the number *.mn.* equals the number *.no.*. But the two remaining numbers I shall demonstrate to be unequal, namely *.kt.* and *.kl.*. Therefore, *.a.* is to *.b.* as *.g.* is to *.d.*; therefore equally will be *.a.* to *.g.* as *.b.* to *.d.*. Therefore, if *.b.* is bigger than *.a.*, *.d.* will be bigger than *.g.*; and if *.b.* is smaller than *.a.*, then also *.d.* will be smaller than *.g.*. Therefore, the numbers *.a.* and *.g.* both are smaller or both are bigger than the numbers *.b.* and *.d.*; equality is indeed not possible, because if they were equal, the numbers *.a.*, *.b.*, *.g.*, *.d.* would not be distinct.

Let, therefore, *.a.* and *.g.* be the smaller numbers; therefore, that made from their product, namely *.tk.*, is smaller than the product made from *.d.* and *.b.*, that is than *.k.*. And *.a.* is to *.g.* as the square of *.a.* is to the product of *.a.* and *.g.*, that is, to the number *.tk.*. Again, *.a.* is to *.g.* as *.b.* is to *.d.*. But *.b.* is to *.d.* as the square of *.b.* is to the product of *.b.* and *.d.*, that is the number *.kl.*. By equality, therefore, *.a.* is to *.g.* as the square of *.b.* is to the number *.kl.*. But as *.a.* is to *.g.* so is the square of *.a.* to the number *.tk.*. Therefore, by addition and proportionality, *.a.* is to *.g.* as the sum of the squares of *.a.* and *.b.* is to the sum of the two numbers *.tk.* and *.kl.*, that is the number *.e.* to the number *.tl.*. Similarly, it is shown that *.a.* is to *.g.* as *.tl.* is to *.z.*. Therefore, *.e.* is to *.tl.* as *.tl.* is to *.z.*. Therefore, the number *.tl.* is the mean proportional between the numbers *.e.* and *.z.*. Therefore, the square of the number *.tl.* is equal to the product of the numbers *.e.* and *.z.*. But the product of the numbers *.e.* and *.z.* is the number *.cf.*. Therefore, *.cf.* is a square with root *.tl.*. In the above demonstration, the number *.cf.* was shown in another way equal to the sum of the four squares of the numbers *.tk.*, *.kl.*, *.mn.*, *.no.*. I shall demonstrate, in fact, that the square of *.tl.* is equal to the sum of these four squares in the following manner. The square of the number *.tl.*, in fact, is equal to the sum of the two squares *.tk.* and *.kl.* plus twice the product of *.tk.* and *.kl.*. But we demonstrated above that the product of *.tk.* and *.kl.* is equal to the product of *.mn.* and *.no.*. But the product of *.mn.* and *.no.* is made from equal numbers. Therefore, *.mn.* times *.no.* is *.mn.* times itself as well as *.no.* times itself. Therefore, twice *.tk.* times *.kl.* is equal to the sum of the two squares of numbers *.mn.* and *.no.*. Therefore, it is demonstrated that the square of the number *.tl.* is equal to the sum of the squares of *.tk.*, *.kl.*, *.mn.*, *no.*, that is the number *.cf.*, as had to be shown. And because the number *.tk.* is smaller than the number *.kl.*, take from the number *.lk.* the number *.kp.* equal to the number *.tk.*; and as we have said above, the numbers *.mo.* and *.pl.* are found with the sum of their squares equal to the number *.cf.*.

Comments on Proposition 7

To find a square number equal to the sum of two square numbers. Leonardo shows how equations (*1) and (*2) of proposition 6 can be used to solve again this Pythagorean problem considered before in propositions 1, 3, 5.

$$(a^2 + b^2)(g^2 + d^2) = [(a)(g) + (b)(d)]^2 + [(b)(g) - (a)(d)]^2. \quad (*1)$$

$$(a^2 + b^2)(g^2 + d^2) = [(a)(d) + (b)(g)]^2 + [(b)(d) - (a)(g)]^2. \quad (*2)$$

By setting the two right sides of the equations equal and by choosing a, b, g, d proportional so that $(b)(g) - (a)(d) = 0$ and $(a)(g) + (b)(d) \neq 0$, he obtains this equation.

$$[(a)(g) + (b)(d)]^2 = [(a)(d) + (b)(g)]^2 + [(b)(d) - (a)(g)]^2.$$

For the numerical example, we choose $a = 3, b = 4, g = 6, d = 8$, which are proportional and make $(b)(g) - (a)(d) = 0$.

$$[(3)(6) + (4)(8)]^2 = [(3)(8) + (4)(6)]^2 + [(4)(8) - (3)(6)]^2.$$

$$50^2 = 48^2 + 14^2.$$

Leonardo's argument in line segment notation is now summarized. a, b, g, d are given proportional numbers so that $(a)(d) = (b)(g)$.

$$(a)^2 + (b)^2 = e. \qquad (g)^2 + (d)^2 = z.$$

$$tk = (a)(g). \qquad kl = (b)(d). \qquad mn = (a)(d). \qquad no = (b)(g).$$

By proportionality, $(a)(d) = (b)(g)$, $mn = no$.

$$cf = (e)(z) = (a^2 + b^2)(g^2 + d^2).$$

$tk \neq kl$, because if $b > a$, then $d > g$;

$$(b)(d) > (a)(g). \qquad kl > tk.$$

If $a < b$, then $d < g$;

$$(b)(d) < (a)(g). \qquad kl < tk.$$

Assume $tk < kl$.

$$a/g = a^2/(a)(g) = a^2/tk.$$

$$a/g = b/d = b^2/(b)(d) = b^2/kl.$$

$$a/g = (a^2 + b^2)/(tk + kl) = e/tl;$$

Similarly, $a/g = tl/z$.

$$e/tl = tl/z. \qquad (e)(z) = tl^2.$$

$$cf = (e)(z) = tl^2 = [(a)(g) + (b)(d)]^2.$$

Note $tl = tk + kl = (a)(g) + (b)(d)$.

$$\begin{aligned}(tl)^2 &= (tk + kl)^2 = (tk)^2 + 2(tk)(kl) + (kl)^2 \\ &= (tk)^2 + 2(mn)(no) + (kl)^2 \\ &= (tk)^2 + (mn)^2 + (no)^2 + (kl)^2. \qquad \text{Note } mn = no. \\ &= cf \\ &= (mn + no)^2 + (kl - tk)^2 = (mo + pl)^2 \\ &= [(a)(d) + (b)(g)]^2 + [(b)(d) - (a)(g)]^2.\end{aligned}$$

Proposition 8

Two squares can again be found whose sum will be the square of the sum of the squares of any two given numbers.

For example, let there be given any two numbers $.a.$ and $.b.$. Let, however, $.b.$ be the bigger; and subtract the square of the number $.a.$ from the square of the number $.b.$, and the difference will be the root of one of the sought squares. Next, twice the product of $.a.$ and $.b.$ is taken, which will be likewise the root of the other square; that was proved in the immediately preceding demonstration. Thus, I put $.g.$ to $.d.$ in the same proportion as $.a.$ to $.b.$. Let $.g.$ be equal to $.a.$ and $.d.$ equal to $.b.$,; and the product of $.a.$ with itself will equal the product of $.a.$ with $.g.$, and this makes $.tk.$; and the product of $.b.$ with itself equals the product of $.b.$ with $.d.$, namely $.kl.$. Then if $.tk.$, that is $.kp.$, is subtracted from $.kl.$, the difference will be $.pl.$, which is one of the roots. Likewise, twice the product of $.a.$ and $.b.$ is equal to $.a.$ times $.d.$ plus $.g.$ times $.b.$, namely the number $.mo.$, which is the other root.

Comments on Proposition 8

Two squares can be found whose sum will be the square of the sum of the squares of any two given numbers. From proposition 7 we have the equation

$$(a^2 + b^2)(g^2 + d^2) = [(a)(d) + (b)(g)]^2 + [(b)(d) - (a)(g)]^2.$$

Assuming $a < b$ and setting $a = g$ and $b = d$ (implying $a/b = g/d$), we have

$$[a^2 + b^2]^2 = [2(a)(b)]^2 + [b^2 - a^2]^2 = mo^2 + pl^2.$$

For a numerical example, let $a = 3$, $b = 5$, $g = 3$, $d = 5$.

$$(3^2 + 5^2)^2 = [2(3)(5)]^2 + [5^2 - 3^2]^2.$$

$$34^2 = 30^2 + 16^2.$$

Proposition 9

Find two numbers which have the sum of their squares equal to a nonsquare number which is itself the sum of the squares of two given numbers.

Let the two given numbers be $.g.$ and $.d.$ and the sum of their squares be the number $.z.$, not a square. I wish to find two other numbers which have the sum of their squares equal to the number $.z.$.

a	g
b	d
e	z
f	
i	
p	q

Take two numbers $.a.$ and $.b.$, which have the sum of their squares equal to the square number $.e.$ with the ratio of $.a.$ to $.b.$ not as $.g.$ to $.d.$. The number $.i.$ results from the multiplication of $.e.$ and $.z.$. Two numbers $.p.$ and $.q.$ are taken, which have the sum of their squares equal to $.i.$. Let $.p.$ and $.q.$ be represented by the line segments $.kl.$ and $.lm.$, making a right angle, namely the angle $.klm.$ connected by the segment $.km.$.

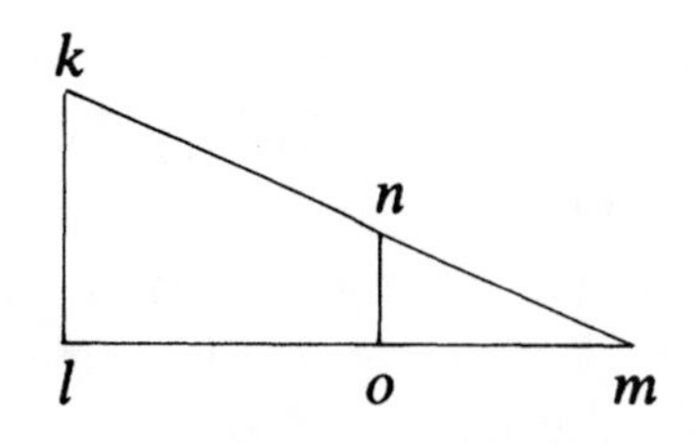

Therefore, .*km*. will be the root of the number .*i*.; and from .*km*. is taken .*mn*., which equals the root of the number .*z*.; and .*no*. is drawn so that .*no*. makes a right angle with .*om*.. The sum of the squares of .*no*. and .*om*. equals the number .*z*.. Hence, the square of .*km*. is equal to the number .*i*.; and the number .*i*. is the product of .*e*. and .*z*.; therefore, if we multiply the root of the number .*e*. by the root of the number .*z*., we have the root of the number .*i*., that is .*mk*.. And since the root of the number .*e*. is known, the quotient of this same root by unity is the quotient of the root of .*i*., namely .*mk*., by the root of .*z*., that is .*mn*.. Let, therefore, .*f*. be the root of the number .*e*.. Because the unity is to the number .*f*. as .*mn*. is to .*km*., and .*mn*. is to .*mk*. as .*no*. is to .*kl*., and .*om*. is to .*lm*., it follows that the unity is to the number .*f*. as .*no*. is to .*kl*. and .*mo*. is to .*ml*.. Therefore, if we divide .*kl*. by the number .*f*., there will result the number .*no*.. Similarly, if we divide .*ml*. by .*f*., there will result .*om*.. Therefore, two numbers .*no*. and .*om*. are found which have the sum of their squares equal to a nonsquare number .*mn*., that is the number .*z*.; this .*z*. is the sum of the squares of the numbers .*g*. and .*d*., which had to be shown.

And here it is shown with numbers. Let the number .*g*. be 4, and the number .*d*. be 5; therefore, the sum of their squares, namely .*z*., is 41. Let the chosen number .*a*. be 3 and .*b*. be 4, which make the sum of the squares be 25, namely, the number .*e*.. From the product indeed of .*e*. and .*z*., namely 25 and 41, arises 1025; and it is possible to find two pairs of other numbers which have the sum of their squares equal to 1025, one of which is 32 and 1, and the other is 31 and 8. Let, therefore, *.kl.* be 32 or 31 and *.lm.* be 1 or 8; and the root of 25, namely the number *.f.*, is taken and each number *.kl.* and *.lm.* is divided by it, and we shall have *.no.* and *.om.*; namely if *.kl.* is 32 and *.lm.* is 1, then *.no.* will be $6\frac{2}{5}$ and *.om.* will be only $\frac{1}{5}$. Therefore, two numbers are found, namely, $6\frac{2}{5}$ and $\frac{1}{5}$, with the sum of their squares equal to 41, that is the number *.z.*. And if *.kl.* is 31 and *.lm.* is 8, *.no.* will be $6\frac{1}{5}$ and *.om.* will be $1\frac{3}{5}$, and thus two other numbers are found which have the sum of their squares likewise 41. This is what had to be done.

Comments on Proposition 9

Find two numbers which have the sum of their squares equal to a nonsquare number which is itself the sum of the squares of two given numbers. g and d are the given numbers with squares summing to z, a nonsquare number. a and b are numbers with squares summing to e, which is square and also $a/b \neq g/d$.

$$g^2 + d^2 = z. \qquad a^2 + b^2 = e.$$

$$i = (e)(z) = (a^2 + b^2)(g^2 + d^2).$$

Let $i = p^2 + q^2$. Such a p and q can be found using equation (*1) of proposition 6, for example. Denote p with the line segment kl and q with the line segment lm, the two segments set at right angles.

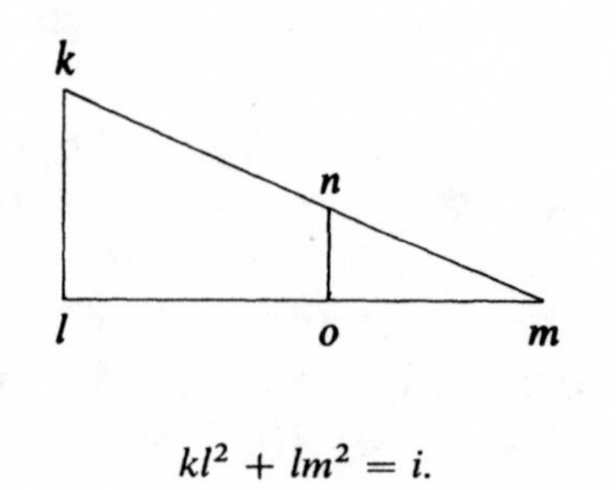

$$kl^2 + lm^2 = i.$$

Line segment *nm* is equal to the square root of *z*.

$$no^2 + om^2 = z.$$

$$(e)^{1/2}(z)^{1/2} = (i)^{1/2} = mk. \qquad (e)^{1/2} = (i)^{1/2}/(z)^{1/2} = f.$$

$$(mn)/(mk) = (no)/(kl) = (om)/(lm).$$

$$1/f = (mn)/(km) = (no)/(kl) = (mo)/(ml).$$

$$(kl)/f = no. \qquad (ml)/f = om.$$

$$(no)^2 + (om)^2 = [(kl)/f]^2 + [(ml)/f]^2 = i^2/f^2 = z.$$

Suppose $g = 4$, $d = 5$, $z = 41$, and $a = 3$, $b = 4$.

$$4^2 + 5^2 = 41. \qquad 3^2 + 4^2 = 5^2.$$

$$(3^2 + 4^2)(4^2 + 5^2) = (25)(41) = 1025.$$

Two squares which sum to 1025 are obtained from

$$p = 32, q = 1, \quad \text{and} \quad p = 31, q = 8;$$

$$p = kl, q = ml, \quad \text{and} \quad p = kl, q = ml.$$

$$(32)^2 + (1)^2 = 1025. \quad (31)^2 + (8)^2 = 1025.$$

$$f = 5.$$

$$(no)^2 + (om)^2 = [(kl)/f]^2 + [(ml)/f]^2$$

$$= (32/5)^2 + (1/5)^2 = [32^2 + 1^2]/5^2 = 1025/25 = 41.$$

$$= (31/5)^2 + (8/5)^2 = [31^2 + 8^2]/5^2 = 1025/25 = 41.$$

The solutions are $6\frac{2}{5}$, $\frac{1}{5}$ and $6\frac{1}{5}$, $1\frac{3}{5}$.

Proposition 10

If, beginning with the unity, a number of consecutive numbers, both even and odd numbers, are taken in order, then the triple product of the last number and the number following it and the sum of the two, is equal to six times the sum of the squares of all the numbers, namely from the unity to the last.

Beginning with the unity *.ab.*, the consecutive numbers both even and odd, *.bg.*, *.gd.*, *.de.*, *.ez.* are taken, and let *.zi.* be the number following the number *.ez.* in order, that is it plus one. I say the product of the numbers *.ez.* and *.zi.* and *.ei.*, that is *.ez.*, *.zi.* and the sum of the two, namely *.ei.*, equals six times the sum of the squares of all the numbers *.ab.*, *.bg.* *.gd.*, *.de.*, *.ez.*. The number *.zt.*, equal to the number *.ez.*, is subtracted from the number *.zi.*, leaving the

a b g d e z k t i

difference *.ti.* equal to one. Again, *.kz.*, equal to the number *.de.*, is subtracted from *.zt.*, leaving the difference *.kt.* equal to 1, by which the number *.ez.* exceeds the number *.de.*. Certainly, the number *.zt.* equals the number *.ez.*. Therefore, number *.ki.* will be twice one, namely 2. Therefore, the triple product of the numbers *.ez.*, *.zk.* and *.ek.* equals the triple product of the numbers *.ze.*, *.ed.*, *.dz.*. But the triple product of the numbers *.ez.*, *.zk.*, *.ek.* plus the triple product of the numbers *.ez.*, *.zk.*, *.ki.* plus the triple product of the numbers *.ez.*, *.ki.*, *.ei.* are equal to the product of the numbers *.ez.*, *.zi.*, *.ei.*. We shall demonstrate, in fact, that the product of *.ez.*, *.zk.*, *.ki.*, plus the product of *.ez.*, *.ki.*, *.ei.*, are equal to six times|the square of the number *.ez.*. I let, then, the number *.ez.* be the root. Therefore, *.zk.* will be the root less one. Therefore, the total *.ei.* is equal to twice the root plus one. The product, in fact, of *.ez.* and *.zk.* is the square less the root. The product, in fact, of this square less the root by the number *.ki.*, namely 2, makes twice the square less twice the root. Therefore, the product of the numbers *.ez.*, *.zk.*, *.ki.* is equal to twice the square of the number *.ez.* less two times the root *.ez.*. Likewise, from *.ez.* times *.ki.* results twice the root, which multiplied by twice the root plus one, namely the number *.ei.*, makes four times the square plus twice the root. Therefore, the triple product of *.ez.*, *.ki.*, *.ei.* is equal to four times the square of the number *.ez.* plus twice the root *.ez.*. Therefore, the aforementioned twice the square less twice the root added to four times the square plus twice the root yields six times the square of number *.ez.*. Therefore, the triple product of the numbers *.ez.*, *.zi.*, *.ei.* is equal to the triple product of the numbers *.ez.*, *.zk.*, *.ek.* plus six times the square of *.ez.*. But the triple product of the numbers *.ez.*, *.zk.*, *.ek.* is equal to the triple product of *.de.*, *.ez.*, *.dz.*. Therefore, the triple product of the numbers *.ez.*, *.zi.*, *.ei.* is equal to the triple product of the numbers *.de.*, *.ez.*, *.dz.* plus six times the square of the number *.ez.*. Similarly, it is shown that the triple product of *.de.*, *.ez.*, *.dz.* is equal to the triple product of *.gd.*, *.de.*, *.ge.* plus six times the square of the number *.de.*. Therefore, the triple product of the numbers *.ez.*, *.zi.*, *.ei.* is equal to the triple product of the numbers *.gd.*, *.de.*, *.ge.* plus six times the sum of the squares of the numbers *.de.* and *.ez.*. It is again shown that the triple product of the numbers *.gd.*, *.de.*, *.ge.* is equal to the triple product of the numbers *.bg.*, *.gd.*, *.bd.* plus six times the square of the number *.gd.*. Therefore, the triple product of the numbers *.ez.*, *.zi.*, *.ei.* is equal to the triple product of the numbers *.bg.*, *.gd.*, *.db.* plus six times the sum of the squares of the numbers *.gd.*, *.de.*, *.ez.*.

These aforementioned things disposed of, it is similarly shown that the product of the numbers *.bg.*, *.gd.*, *.bd.* is equal to the sum of the triple product of the unity *.ab.*, *.bg.* and *.ag.*, and six times the square of the number *.bg.*. Therefore, the triple product of the numbers *.ez.*, *.zi.*, *.ei.* is equal to the triple product of the unity *.ab.*, the sum of *.bg.* and *.ag.*, and six times the sum of the squares of the numbers *.bg.*, *.gd.*, *.de.*, *.ef.*. But the triple product of the unity *.ab.* and the numbers *.bg.* and *.ag.* is equal to six times the square of *.ab.*; for *.bg.* is 2 and *.ag.* is 3. Therefore, the triple product of the numbers *.ez.*, *.zi.*, *.ei.* is equal to six times the sum of the squares of the unity *.ab.* and the numbers *.bg.*, *.gd.*, *.de.*, *.ez.*. This is what had to be shown. There is indeed another way by which we can discover the same thing; this is shown in what follows.

Comments on Proposition 10

In modern notation, proposition 10 is equivalent to the equation

$$6(1^2 + 2^2 + 3^2 + \cdots + n^2) = n(n + 1)(2n + 1).$$

Leonardo first proves the equation

$$n(n + 1)(2n + 1) = (n - 1)n(2n - 1) + 6n^2.$$

This equation is the essential step in a proof by mathematical induction. Leonardo also proves other equations for $n - 1, n - 2, \ldots, 3, 2, 1$. Here is the list.

$$\begin{array}{lcl} n(n+1)(2n+1) & -\,(n-1)n(2n-1) & = 6n^2. \\ (n-1)n(2n-1) & -\,(n-2)(n-1)(2n-3) & = 6(n-1)^2. \\ (n-2)(n-1)(2n-3) & -\,(n-3)(n-2)(2n-5) & = 6(n-2)^2. \\ & \vdots & \vdots \\ (3)(4)(7) & -\quad (2)(3)(5) & = 6(3)^2. \\ (2)(3)(5) & -\quad (1)(2)(3) & = 6(2)^2. \\ (1)(1+1)(2+1) & & = 6(1)^2. \end{array}$$

The equations are then summed. On the left, only the first term remains after all the cancellation.

$$n(n + 1)(2n + 1) = 6(1^2 + 2^2 + 3^2 + \cdots + n^2).$$

We now show the argument with Leonardo's notation. *ab*, *bg*, *gd*, *de*, *ez*, *zi* are consecutive whole numbers beginning with the unity. *kt* and *ti* are both equal to the unity.

a b g d e z k t i

Leonardo demonstrates that

$$n(n + 1)(2n + 1) = 6(1^2 + 2^2 + 3^2 + \cdots + n^2)$$

in the form

$$(ez)(zi)(ei) = 6(ab^2 + bg^2 + gd^2 + de^2 + ez^2).$$

$ti = 1$ implies $zt = ez$. $kt = 1, ki = 2$, imply $zk = de$.

$(ez)(zk)(ek) = (ez)(de)(dz)$.

Leonardo argues as follows:

$zk = ez - 1, \quad ei = 2ez + 1. \quad ki = 2.$

$(ez)(zk) = (ez)^2 - ez. \quad (ez)(zk)(ki) = 2(ez)^2 - 2(ez).$

$(ez)(ki) = 2(ez). \quad (ez)(ki)(ei) = 2(ez)(2ez + 1) = 4(ez)^2 + 2(ez).$

$(ez)(zk)(ki) + (ez)(ki)(ei) = 2(ez)^2 - 2(ez) + 4(ez)^2 + 2(ez) = 6(ez)^2.$

To each side of the equation

$$(ez)(zk)(ki) + (ez)(ki)(ei) = 6(ez)^2.$$

Leonardo adds $(ez)(zk)(ek)$.

$$(ez)(zk)(ek) + (ez)(zk)(ki) + (ez)(ki)(ei) = (ez)(zk)(ek) + 6(ez)^2.$$

But $ek + ki = ei$.

$$(ez)(zk)(ei) + (ez)(ki)(ei) = (ez)(zk)(ek) + 6(ez)^2.$$

And $zk + ki = zi$.

$$(ez)(zi)(ei) = (ez)(zk)(ek) + 6(ez)^2.$$

But $(ez)(zk)(ek) = (de)(ez)(dz)$. This gives the inductive equation

$$(ez)(zi)(ei) = (ez)(zk)(ek) + 6(ez)^2.$$

Similarly, Leonardo accumulates the formulas for the squares that come before.

$$(de)(ez)(dz) = (gd)(de)(ge) + 6(de)^2.$$

$$(gd)(de)(ge) = (bg)(gd)(bd) + 6(gd)^2.$$

$$(bg)(gd)(bd) = (ab)(bg)(ag) + 6(bg)^2.$$

$$(ab)(bg)(ag) = (1)(2)(3) = 6(ab)^2.$$

The results are combined to give the desired equation.

$$(ez)(zi)(ei) = 6[ab^2 + bg^2 + gd^2 + de^2 + ez^2].$$

A slightly different way of finding the equation containing the square of ez is to make the following combinations of line segments.

$$(ez)(zk)(ki) + (ez)(ki)(ei) = (ez)(ki)[zk + ei]$$
$$= (ez)(ki)[de + ei] = (ez)(ki)(di) = (ez)(ki)[de + ez + zt + ti]$$
$$= (ez)(ki)[ez + ez + ez] = 6(ez)^2.$$

Proposition 11

If, beginning with the unity, a number of consecutive odd numbers are taken in order, then the triple product of the last number and the odd number following it and their sum is equal to twelve times the sum of all the squares of the odd numbers from the unity to the last odd number

Beginning with the unity *.ab.* let, in fact, *.bg.*, *.gd.*, *.de.* be consecutive odd numbers and let the odd number following *.de.* be *.ez.*. I say, in fact, that the triple product of the numbers *.de.*, *.ez.* and their sum *.dz.* equals twelve times the sum of the squares of the unity *.ab.* and the odd numbers *.bg.*, *.gd.*, *.de.*. The number *.ei.*,

a b g d e i z

in fact, is taken equal to the number *.de.* on the segment *.ez.*; therefore, *.iz.* will be two. I shall show first that the triple product of the numbers *.gd.*, *.de.* and their sum *.ge.* added to twelve times the square of the number *.de.* is equal to the triple product of the numbers *.de.*, *ez.*, *.dz.*. Therefore, let the number *.de.* be the root; then the number *.gd.* will be the root minus two, and the sum *.ge.* will be twice the root less two. Therefore, from the multiplication of *.gd.* and *.de.* results the square less two roots which, when multiplied by *.ge.*, that is twice the root less two, yields twice the cube of the number plus four times the root less six times the square; to which if is added twelve times the square of the root *.de.*, will become twice the cube plus six times the square plus four times the root. Again, because *.de.* is a root the number *.ei.* will similarly be a root. Therefore, the total *.ez.* will be the root plus two, which is *.iz.*, and the sum *.dz.* will be twice the root plus two. In fact, from the multiplication of *.de.* with *.ez.* results a square plus twice the root. And, in fact, out of the multiplication by the number *.dz.*, that is twice the root plus two, results similarly twice the cube plus six times the square plus four times the root. Therefore, it is shown that the triple product of the numbers *.gd.*, *.de.*, *.ge.* plus twelve times the square of the number *.de.* are equal to the triple product of the numbers *.de.*, *.ez.*, *.dz.*. By the same method it is shown that the triple product of the numbers *.bg.*, *.gd.*, *.bd.* plus twelve times the square of the number *.gd.* equal the triple product of the numbers *.gd.*, *.de.*, *.ge.*. Therefore, the triple product of the numbers *.de.*, *.ez.*, *.dz.* is equal to the product of the numbers *.bg.*, *.gd.*, *.bd.* plus twelve times the sum of the squares of the numbers *.gd.* and *.de.*. Again, with the aforementioned things ordered, the triple product of the numbers *.bg.*, *.gd.*, *bd.* will be shown equal to the triple product of the unity *.ab.* and the numbers *.bg.* and *.ag.* plus twelve times the square of the number *.bg.*. But the triple product of the unity *.ab.* and the numbers *.bg.* and *.ag.* is twelve times the square of the unity, *.ab.*; the number *.bg.* is three and also the number *.ag.* is four. Therefore, the triple product of the numbers *.de.*, *.ez.*, *.dz.* is equal to twelve times the sum of all the squares of the given numbers, namely the unity *.ab.* and the numbers *.bg.*, *.gd.*, *de.*. This is what had to be shown.

By a similar method, if beginning with the number two, consecutive even numbers are taken in order, the triple product of the last of them, the number following it, and the sum of the two will be found equal to twelve times the sum of all the squares of the given even numbers. By the same way and method again, if consecutive multiples of three are taken in ascending order beginning with three, the triple product of the last of them, the number following it, and the sum of both, equals eighteen times the sum of all the squares of the given numbers ascending by three. And then they ascend by twos and are even; then this last triple product is equal to twelve times the sum of all the squares of the given numbers. And when they ascend by units as with consecutive numbers, then the aforementioned triple product is equal to six times the sum of the squares of the given numbers. We demonstrated this and it is written above. It is understood that the sum of the squares of ascending multiples of four will be four times six times the aforementioned product, and that the sum of the squares of ascending multiples of five will be five times six times the triple product, and so on for the rest of the numbers.

Comments on Proposition 11

The formula that Leonardo establishes is

$$(2n-1)(2n+1)4n = 12[1^2+3^2+5^2+\cdots+(2n-1)^2].$$

This equation gives the sum of the squares of consecutive odd numbers. The equation is proven by showing a list of equations and then by adding them together to produce the sum of the squares. The equations are

$$\begin{aligned}
(2n-1)(2n+1)4n &= (2n-3)(2n-1)4(n-1) + 12(2n-1)^2 \\
(2n-3)(2n-1)4(n-1) &= (2n-5)(2n-3)4(n-2) + 12(2n-3)^2. \\
&\ \ \vdots \\
(5)(7)(12) &= (3)(5)(8) + 12(5)^2. \\
(3)(5)(8) &= (1)(3)(4) + 1(3)^2 \\
(1)(3)(4) &= 12(1)^2.
\end{aligned}$$

Now add all the equations together and cancel like terms.

$$(2n - 1)(2n + 1)4n = 12[1^2 + 3^2 + 5^2 + \cdots + (2n - 1)^2].$$

Now we follow Leonardo's argument with line segments. First the inductive formula

$$(gd)(ge)(de) + 12(de)^2 = (de)(ez)(dz)$$

is established.

$$(gd) = (de) - 2. \qquad (ge) = 2(de) - 2.$$

$$(dg)(ge) = (de)[(de) - 2] = (de)^2 - 2(de).$$

$$(ge)(dg)(de) = [2(de) - 2][(de)^2 - 2(de)]$$

$$= 2(de)^3 + 4(de) - 6(de)^2.$$

Twelve times the square of *de* is added to both sides of the equation.

$$(ge)(dg)(de) + 12(de)^2 = 2(de)^3 + 4(de) + 6(de)^2.$$

Holding this equation, the value of $(dz)(de)(ez)$ is found.

$$(de) = (ei). \qquad (ez) = (de) + 2. \qquad (dz) = 2(de) + 2.$$

$$(de)(ez) = (de)^2 + 2(de).$$

$$(dz)(de)(ez) = 2(de)^3 + 6(de)^2 + 4(de).$$

Therefore,

$$(dz)(de)(ez) = (ge)(dg)(de) + 12(de)^2.$$

The next square formula is

$$(bg)(gd)(bd) + 12(gd)^2 = (gd)(de)(ge).$$

Therefore,

$$(de)(ez)(dz) = (bg)(gd)(bd) + 12[(de)^2 + (gd)^2].$$

Also,

$$(bg)(gd)(bd) = (ab)(bg)(ag) + 12(bg)^2.$$

But

$$(ab)(bg)(ag) = (1)(3)(4) = 12(1)^2 = 12(ab)^2.$$

$$(de)(ez)(dz) = 12[(de)^2 + (gd)^2 + (bg)^2 + (ab)^2].$$

There are also the formulas

$$12[2^2 + 4^2 + 6^2 + \cdots + (2n)^2] = (2n)(2n + 2)(4n + 2);$$

$$18[3^2 + 6^2 + 9^2 + \cdots + (3n)^2] = (3n)(3n + 3)(6n + 3);$$

$$24[4^2 + 8^2 + 12^2 + \cdots + (4n)]^2 = (4n)(4n + 4)(8n + 4).$$

Leonardo outlines how these may be obtained.

Proposition 12

If two numbers are relatively prime and have an even sum, and if the triple product of the two numbers and their sum is multiplied by the number by which the greater number exceeds the smaller number, there results a number which will be a multiple of twenty-four.

Let two relatively prime numbers $.ab.$ and $.bg.$ be given, the sum of them be the even number $.ag.$, the ratio of the number $.ab.$ to the number $.bg.$ be in lowest terms. Let $.bg.$ be the larger number. From the number $.bg.$ is taken the number $.bd.$ equal to the number $.ab.$.

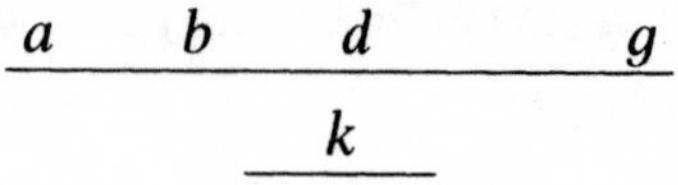

a *b* *d* *g*

k

Therefore, the number $.bg.$ will exceed the number $.ab.$ by the number $.dg.$. I say, in fact, that if the number $.ab.$ is multiplied by the number $.bg.$ and this is multiplied by the number $.ag.$, and this product is multiplied by $.dg.$, there will result a number which is a multiple of twenty-four; dividing by three times eight or by four times six will yield a whole number. The numbers $.ab.$ and $.bg.$ are, in fact, both odd; for if one were odd their sum would not be even. Neither can both be even; for then they would not be relatively prime. Therefore, both numbers $.ab.$ and $.bg.$ are odd. And because the number *.bd.* is equal to the number *.ab.*, the number *.ad.* is therefore twice the number *.ab.*. Therefore, the number *.ad.* is even. Therefore, the difference *.dg.* is even; for if an even number is subtracted from an even number, there remains an even number; and because *.dg.* is even, half of it will be a whole number, even or odd. First, let the half be odd. Half of the number *.ad.*, namely, *.ab.*, is odd. The sum of half of *.ad.* with half of *.dg.*, namely half of the number *.ag.*, is even because the sum of two odd numbers is an even number. Therefore, *.ag.* is even even, that is half of the number *.ag.* is even. The total number *.ag.* will be a multiple of four. Therefore, from the multiplication of *.ag.* and *.gd.* arises a number which is a multiple of eight. But if half the number *.gd.* is even, then *.gd.* will be a multiple of four. Therefore, from the multiplication of *.dg.* and *.ag.* will result a number which is similarly a multiple of eight. Therefore, if the product of *.ag.* and *.bg.* is multiplied by *.ab.*, there will result a number which is a multiple of eight. And because *.ab.* and *.bg.* are odd, one of them is a multiple of three or neither is. Suppose the first. Therefore, from the multiplication of the numbers *.ab.*, *.bg.*, *.ag.* and the number *.dg.* results the number *.k.*, which is a multiple of three and also a multiple of eight. Therefore, *.k.* will be a multiple of twenty-four as we said before. And if neither of the numbers *.ab.* or *.bg.* is a multiple of three, and if each of them is divided by 3, their remainders will be equal or unequal. First, suppose they are equal. Then the number *.gd.* is a multiple of three. Therefore, the aforementioned product multiplied by *.dg.* yields a number which is a multiple of three. But if the remainders upon dividing *.ab.* and *.bg.* by 3 are unequal, then one of the remainders will be 1 and the other will be 2. Therefore, the sum of *.ab.* and *.bg.*, namely the number *.ag.*, will be a multiple of three. Hence the triple product of *.ab.*, *bg.*, *ag.* will be a multiple of three. Therefore, from the multiplication of this triple product with the number *.dg.* results a number which is a multiple of three. And because similarly it is a multiple of eight, it will be therefore a multiple of twenty-four, which had to be shown. And this will again be true if the numbers *.ab.* and *.bg.* are not relatively prime.

And if one of the numbers *.ab.* and *.bg.* is even, the sum of them will be odd; then it will be similarly shown that from the product of the doubles of each of the numbers and their sum and the number *.dg.* will arise a number which will be a multiple of twenty-four whether the numbers are relatively prime or not. This obtained number, namely the multiple of twenty-four, is called congruous.

Comments on Proposition 12

If m and n are relatively prime and both are odd numbers, then $mn(n + m)(n - m)$ is a multiple of 24.

These numbers that have the form $mn(n + m)(n - m)$ are used by Leonardo in Proposition 14; he calls them congruous numbers (*numerus congruus*).

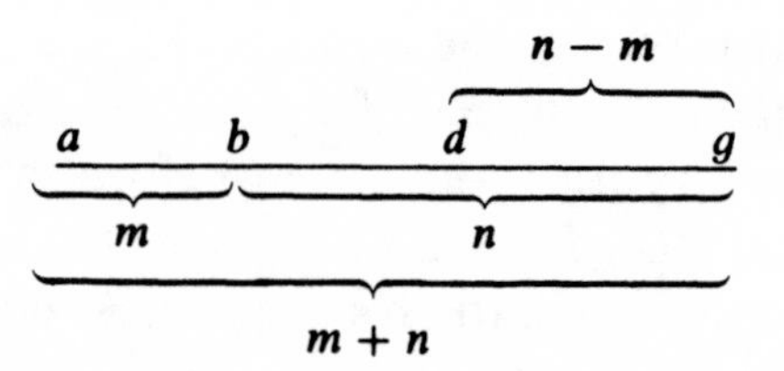

We exhibit the argument first in Leonardo's notation. It must be shown that $(ab)(bg)(ag)(dg)$ is a multiple of 24. Let ab and bg be the given numbers, relatively prime, $bg > ab$, and both odd. Let d be placed so that $ab = bd$. Then $ad = 2(ab)$. $ag = ab + bg$. $dg = ag - ad$. Both ag and dg are even numbers. Therefore, both ag and dg are multiples of two.

Case 1. $\frac{1}{2}(dg)$ is an odd number. $\frac{1}{2}(ad) = ab$ is also given an odd number. Therefore, $\frac{1}{2}(ag) + \frac{1}{2}(ad) = \frac{1}{2}(ag)$ is an even number. $\frac{1}{2}(ag)\frac{1}{2}(dg)$

is an even number. $\frac{1}{4}(ag)(dg)$ is an even number. $(ag)(dg)$ is a multiple of 8.

Case 2. $\frac{1}{2}(dg)$ is an even number. In this case, both $\frac{1}{2}(ad)$ and $\frac{1}{2}(ag)$ are odd numbers. But certainly $\frac{1}{2}(ag)\frac{1}{2}(dg)$ is an even number. $\frac{1}{4}(ag)(dg)$ is an even number. $(ag)(dg)$ is a multiple of 8.

Now *ab* and *bg* have no common factor. Their remainders upon dividing by 3 are either 1 or 2. If they have the same remainder, then their difference *dg* is a multiple of 3. If they have different remainders, then their sum *ag* is a multiple of 3. In either alternative, $(ab)(bg)(ag)(dg)$ is a multiple of 3.

Since $(ab)(bg)(ag)(dg)$ is a multiple of both 8 and 3, it must be a multiple of 24. This is Leonardo's argument.

Rephrased in modern notation, $(ab) = m$, $(bg) = n$, $(ag) = m + n$, $(dg) = n - m$. Both m and n are odd and relatively prime. Both $m + n$ and $n - m$ are even.

Case 1. $(n - m)/2$ is odd. $m + (n - m)/2 = (m + n)/2$ is even. $[(n - m)/2][(m + n)/2]$ is even. $(m + n)(n - m)/4$ is even. $(m + n)(m - n)$ is a multiple of 8.

Case 2. $(n - m)/2$ is even. $[(n - m)/2][(m + n)/]$ is even. $(m + n)(n - m)/4$ is even. $(m + n)(n - m)$ is a multiple of 8.

In either case, $mn(m + n)(n - m)$ is a multiple of 8. Then m and n have no common factor. Upon dividing by 3, the remainders for m and n must be either 1 or 2. If both are 1 or both are 2, then the difference $n - m$ has a remainder of 0 upon dividing by 3. This is to say, $n - m$ is a multiple of 3. $m = 3p + 1$ and $n = 3q + 1$ for some p and q imply $n - m = 3q - 3p = 3(q - p)$. Or if $m = 3p + 2$ and $n = 3q + 2$, then $n - m = (3q + 2) - (3p + 2) = 3(q - p)$, a multiple of 3. If $n - m$ is a multiple of 3, then so is $mn(m + n)(n - m)$. On the other hand, if the remainders are different, one remainder is 1 and the other is 2, then it is the sum of m and n that is a multiple of 3. $m + n = (3p + 1) + (3q + 2) = 3p + 3q + 3 = 3(p + q + 1)$ or $m + n = (3p + 2) + (3q + 1) = 3(p + q + 1)$. In either case, $m + n$ is a multiple of 3 and so is $mn(m + n)(n - m)$. If case m and n are relatively prime and one of the two numbers is even, then $(2m)(2n)(m + n)(n - m)$ is a multiple of 24. This case is mentioned but not proven by Leonardo.

Either $2m$ or $2n$ is a multiple of 4, and therefore $(2m)(2n)$ is a multiple of 8. If either m or n is a multiple of 3, then $(2m)(2n)(m + n)(n - m)$ is a multiple of 3. If neither m nor n is a multiple of 3, then we argue as before to show either $m + n$ or $n - m$ is a multiple of 3. In either case, $(2m)(2n)(m + n)(n - m)$ is a multiple of 3 and 8 and therefore of 24.

Now in case the two numbers m and n are not relatively prime, the result still holds. Suppose that k is the greatest common divisor of m and n. $m = kM$ and $n = kN$, and M and N are relatively prime. If both M and N are odd, then $MN(N + M)(N - M)$ is a multiple of 24. Then

$$\begin{aligned} mn(n + m)(n - m) &= (kM)(kN)(kM + kN)(kN - kM) \\ &= k^4 MN(M + N)(N - M) \end{aligned}$$

is a multiple of 24. In case either M or N is even, then

$$(2m)(2n)(m + n)(n - m) = k^4(2m)(2n)(m + n)(n - m)$$

is a multiple of 24.

Proposition 13

If about some given number are located some smaller and larger numbers and if the number of smaller numbers equals the number of larger numbers, and if each of the larger numbers exceeds the given number by the same as the given number exceeds a smaller number, then the sum of all the smaller and larger numbers will be the product of the number of located numbers and the given number.

Let numbers $.a.$, $.b.$, $.g.$, $.e.$, $.z.$, $.i.$ be taken about the

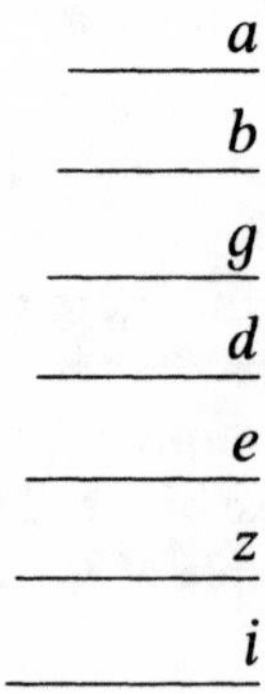

a

b

g

d

e

z

i

number $.d.$, and let $.a.$ be the smallest of them and $.i.$ be the largest. Let the number $.i.$ exceed the number $.d.$ by as much as the number $.d.$ exceeds the number $.a.$. Similarly, let the number $.z.$ exceed the number $.d.$ by as much as the number $.d.$ exceeds the number $.b.$. Again, let the number $.e.$ exceed the number $.d.$ by as much as the number $.d.$ exceeds the number $.g.$. I say that if the number $.d.$ is multiplied by the number of numbers $.a.$, $.b.$, $.g.$, $.e.$, $.z.$, $.i.$, the product will be the sum of all these numbers, which is proven thus. I diminish, in fact, the number $.i.$ by as much as it exceeds $.d.$ and add the difference to $.a.$; each of the numbers obtained from $.i.$ and $.a.$ will be equal to the number $.d.$. I do the same to the numbers $.z.$, $.b.$, and $.e.$, $.g.$, and each of them will be altered to the number $.d.$. Therefore, the sum of all the numbers $.a.$, $.b.$, $.g.$, $.e.$, $.z.$, $.i.$ is equal to the number of numbers $.a.$, $.b.$, $.g.$, $.e.$, $.z.$, $.i.$ times the number $.d.$, which had to be shown.

Comments on Proposition 13

The numbers a, i, b, z, g, e are symmetrically placed in pairs about the central number d.

$$i - d = d - a. \qquad z - d = d - b. \qquad e - d = d - g.$$

This yields

$$a + i = 2d. \qquad b + z = 2d. \qquad g + e = 2d.$$

Summing both columns gives

$$a + b + g + e + z + i = 6d.$$

The average of a, i, b, z, g, e is d.

Proposition 14

Find a number which added to a square number and subtracted from a square number yields always a square number.

And thus must be found three squares and a number so that the number added to the smallest square makes the second square, and the same number added to the second square makes the third square, which is the greatest. And thus adding this number to, and subtracting it from, the second square yields always a square. Because all square numbers result from sums of consecutive odd numbers, the smallest square number will be the sum of some consecutive odd numbers beginning with the unity. To that square it is proposed to add a number, and make a second square. This square again is the sum of a number of consecutive odd numbers beginning with the unity. If the same number that is called congruous because it makes the squares congruent is added to the second square, the sum is the greatest square, which will similarly also be made from the sum of some consecutive odd numbers beginning with the unity. In the total number of odd numbers there are those whose sum makes the first square, and another set whose sum makes the congruous number and another set whose sum makes the same congruous number. Therefore, the set of odd numbers is divided into the three aforementioned parts. But the number of odd numbers making the first congruous number is in some ratio with the number making the second; in fact, there are more odd numbers making the first congruous number than odd numbers making the second congruous number, because the smaller numbers come first in the increasing order; therefore, in the first quantity are the smaller odd numbers and in the other are the larger. Therefore, we shall manage to show by this method that the congruous number can be found with some given possible proportion.

Take two arbitrary numbers $.ab.$ and $.bg.$; the sum of them is $.ag.$, and let $.bg.$ be bigger than $.ab.$ by the number $.dg.$. If, in fact, the ratio of the number $.bg.$ to the number $.ab.$ is smaller than the ratio of the number $.ag.$ to the number .*dg*., then it will be possible to find the congruous number out of a quantity of odd numbers having a ratio to a quantity of following odd numbers, namely the same as that of the number .*gb*. to the number .*ab*.. And if the ratio of the number .*bg*. to the number .*ba*. is greater than the number .*ag*. to the number .*dg*., then it will be impossible to find two quantities of consecutive odd numbers in the ratio of the number .*gb*. to the number .*ab*.. Then it will be possible to find two quantities of consecutive odd numbers in the ratio of the number .*ag*. to the number .*dg*.. And the congruous number will result from the sums of each of these quantities.

First, let the ratio, that of the number .*bg*. to the number .*ab*., be less than that of the number .*ag*. to the number .*dg*.; therefore, there must be found two quantities of consecutive odd numbers in the ratio of the number .*bg*. to the number .*ab*., and so that the sum of the numbers in each quantity is the same. The sum of the numbers .*ab*. and .*bg*., namely .*ag*., is indeed even or odd. First, let it be even. The number .*ad*., namely twice .*ab*., is even. Whence the difference .*dg*. is necessarily even. From the multiplication, in fact, of .*dg*. and .*bg*. results the number .*ez*., and from .*dg*. times .*ab*. results .*zi*.. Therefore, the number .*ez*. is to the number .*zi*. as the number .*gb*. is to the number .*ba*.. And the numbers .*ez*. and .*zi*. are even.

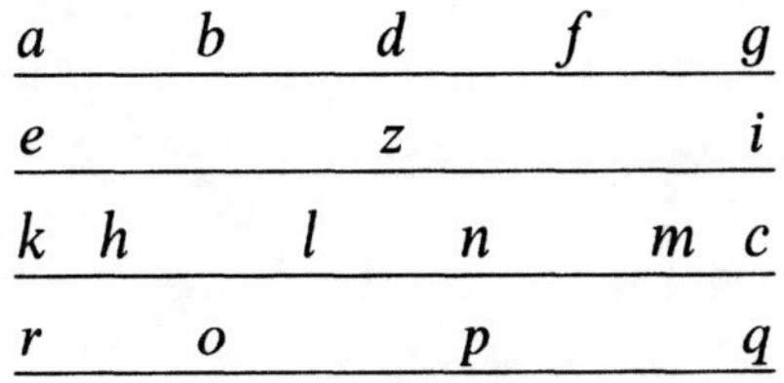

a *b* *d* *f* *g*

e *z* *i*

k *h* *l* *n* *m* *c*

r *o* *p* *q*

Again, from the multiplications of .*ag*. with the numbers .*gb*. and .*ba*. result the even numbers .*km*. and .*kl*.. Therefore, as .*gb*. is to .*ba*., so is .*km*. to .*kl*.. But as .*gb*. is to .*ba*., so is *.ez.* to *.zi.*; therefore, as *.ez.* is to *.zi.* so is *.km.* to *kl.*. Finally, from the multiplication of *.ez.* and *.kl.* results *.op.*; from the multiplication of *.zi.* and *.km.* results *.pq.*. And *.ez.* is to *.zi.* as *.km.* is to *.kl.*. Therefore, that which is the product of *.ez.* and *.kl.* is equal to the product of *.zi.* and *.km.*. Therefore, the number *.op.* is equal to the number *.pq.*. I shall show both of them are congruous numbers.

Because the product of *.ez.* and *.kl.* is *.op.*, there are as many numbers equal to the number *.kl.* in the number *.op.* as there are unities in the number *.ez.*. But in the number *.op.* there are as many consecutive odd numbers about the number *.kl.* as there are numbers equal to *.kl.* in the number *.op.*. Therefore, as many as there are unities in the number *.ez.* as many there are consecutive odd numbers placed about the number *.kl.*, half of them smaller than the number *.kl.*, half of them greater than the number *.kl.*, all of them making a sum equal to *.op.*. Similarly, it is shown that there are as many consecutive odd numbers about the number *.km.* in the number *.pq.* as there are unities in the number *.zi.*. Therefore, the number *.ez.* is to *.zi.* as the number of consecutive odd numbers making the number *.op.* is to the number of consecutive odd numbers making the number *.pq.*, which is shown equal to the number *.op.*. I say, moreover, that both of these quantities of odd numbers are consecutive. For out of the product of *.ag.* and *.bg.* results *.km.*, and *.bg.* is equal to the sum of the numbers *.gd.* and *.db.*. Therefore, from the multiplication of *.ag.* and the sum of *.dg.* and *.db.* results *.km.*. But from the multiplication of *.ag.* and *.bd.*, that is *.ab.*, results *.kl.*. Therefore, the difference *.lm.* is the product of *.ag.* and *.dg.*. But this number, which is the product of *.ag.* and *.dg.*, equals *.gd.* times *.gb.* plus *.gd.* times *.ba.*, that is the sum of *.ez.* and *.zi.*. Therefore, the number *.lm.* equals the number *.ei.*. The number *.ln.*, equal to the number *.ez.*, is taken away from the number *.lm.*. The difference *.nm.* is therefore equal to the difference *.zi.*. Also, it is shown that the number *.kl.* is greater than the number *.ez.*, because the ratio of *.gb.* to *.ba.* is less than the ratio of the numbers *.ag.* to *.gd.*. Therefore, a number having the ratio less than that of the number *.ag.* to the number *.gd.*, having the same ratio to the number *.dg.* as the number *.gb.* to the number *.ba.*; this number is *.af.*. And because *.gb.* to *.ba.* is as *.af.* to *.dg.*, the product of *.af.* and *.ba.* is equal to the product of *.bg.* and *.dg.*, this is the number *.ez.*. But the product of *.ag.* and *.ab.*, this is the number *.kl.*, is greater than the product of *.af.* and *.ab.*, that is the number *.ez.*. From the number *.kl.*, in fact, the number *.lh.* is taken equal to the number *.ln.*, that is *.ez.*, and the number total *.hn.* will be twice the number *.ez.*. Alternatively, to the number *.km.* is added the number *.mc.*, equal to the number *.mn.*, which is *.zi.*. Therefore, the number *.nc.* is twice the number *.zi.*. And because the number *.kl.* is even, if subtracted from it is the number *.lh.*, that is *.ez.*, which is even, there will remain the even number *.kh.*. Similarly, if to *.kl.* is added *.ln.*, the total will be the even *.kn.*. Moreover, the number *.nc.* is even; consequently, the total *.kc.* is even. And because *.kh.* is even, as many unities there are in its half, as many odd numbers there are between the unity and the number *.kh.*, the sum of which make a square number which is the product of half the number *.kh.* with itself. Let this square number be *.ro.*. Similarly, as many unities there are in the number *.hn.*, as many whole numbers there are between the number *.kh.* and the number *.kn.*, an even number of them, because *.hn.* was set even. But half the number *.hn.* is *.ez.*. Therefore, as many unities there are in the number *.ez.*, as many odd numbers there are between the numbers *.kh.* and *.kn.*. The sum of these odd numbers was shown to yield the number *.op.*, because about the number *.kl.* were put half of these numbers between *.kh.* and *.kl.* and the other half between *.kl.* and *.kn.*. Therefore, from the sum of all the odd numbers between the unity and the number *.kn.* results *.rp.*, a square with root half the number *.kn.*.

Again, as many unities there are in the number *.no.*, as many there are of whole numbers between the numbers *.kn.* and *.kc.*, half of which are odd numbers. Therefore, as many unities there are in the number *.mn.*, that is *.zi.*, as many odd numbers there are between the numbers *.kn.* and *.kc.*, and are consecutive with the odd numbers which are between *.kh.* and *.kn.*, as we have said before. But the sum of the odd numbers between *.kn.* and *.kc.*, located about the number *.km.*, is the number *.pq.*. Therefore, the number *.rq.* results from the sum of all the odd numbers between the unity and the number *.kc.*. Therefore, the number *.rq.* is a square and the root of it is half the number *.kc.*. Therefore, each of the numbers *.op.* and *.pq.* is congruous, as we have said before. A number is indeed found, namely *.pq.*, which added to a square, namely to *.rp.*, makes a square number, which is *.rq.*; and when the same number, namely *.pq.*, that is *.op.*, is subtracted from the same square, namely *.rp.*, there remains a square, namely *.ro.*. This is what had to be done.

But to see these things still more clearly, I let the number *.ab.* be 3, also *.bg.* be 5. Then the total *.ag.* will be 8, the difference *.dg.* is 2. The numbers *.ag.* and *.dg.* are certainly even. From the multiplication of *.dg.* and *.bg.* results, in fact, 10. And from the multiplication of *.dg.* and *.ab.* results 6; therefore, *.ez.* is 10 and *.zi.* is 6. Therefore, *.bg.* is to *.ab.* as *.ez.* is to *.zi.*; that is 5 is to 3 as 10 is to 6. And 10 is the number of odd numbers first making the congruous number and 6 is the number of odd numbers secondly making the same congruous number. Again, from the multiplication of *.ag.* by *.gb.* and by *.ba.*, that is 8 by 5 and by 3, results *.km.*, 40, and *.kl.*, 24; the difference is *.lm.*, 16, which is equal to the sum of the numbers *.ez.* and *.zi.*. In fact, the number *.ln.*, equal to the number *.ez.*, namely 10, added to the number *.kl.*, will be *.kn.*, 34. And to *.km.* is added *.nm.*, namely *.zi.*, which is 6, resulting in *.kc.*, 46. And from the number *.kl.* is subtracted *.hl.* equal to *.ln.*, namely 10, leaving *.kh.*, 14. And the multiplication of *.ez.*, that is *.ln.*, by *.kl.*, namely 10 by 24, yields *.po.*, 240, which is a congruous number, and is the sum of the ten odd numbers placed about 24, which are between the numbers *.kh.* and *.kn..* that is between 14 and 34. Again, from the multiplication of *.zi.* by *.km.*, that is 6 by 40, results 240, that is *.pq.*, which is the sum of the six odd numbers existing about 40, which are between *.kn.* and *.kc.*, that is between 34 and 46. In fact, from the odd numbers which are

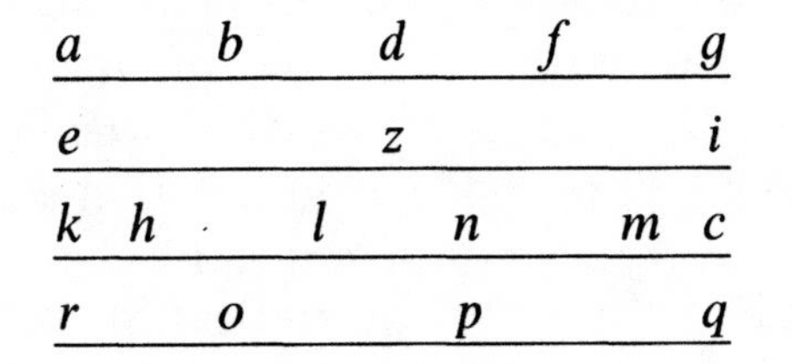

between one and the number *.kh.*, namely 14, results the sum *.ro.* which is 49, because between one and 14 are seven odd numbers placed about seven; therefore, 7 times 7 yields the sum of the odd numbers which are between one and 14. Therefore, *.ro.* is a square. In fact, from the odd numbers which are between *.kh.* and *.hn.* results the sum *.op..* Therefore, from the set of odd numbers which are between one and *.kn.*, namely 34, results the sum *.rp..* And there are 17 of these odd numbers which have sum the product 17 times itself. Therefore, *.rp.* is a square, and is 289. For adding *.ro.* with *.op.*, namely 49 with 240, yields 289, which has root 17, namely half *.kn..* Again, from the set of odd numbers which are between *.kn.* and *.kc.* results the sum *.pq.*, which is equal to the number *.op..* It is certainly 240.

Therefore, from the set of odd numbers which are between one and 46, results the sum *.rq.*, which is a square, with root half the number *.kc.*, namely 23. For adding *.pq.* to *.rp.*, namely 240 to 289, yields 529, which has root 23.

But if the ratio of *.gb.* to *.ba.* is greater than the ratio of *.ag.* to *.gd.*, and if the number *.ag.* is again even, I say it is possible to find two sets of consecutive odd numbers in the ratio of the number *.ag.* to the number *.gd.* making congruous numbers. And to see it clearly, let *.ab.* be 1, and *.bg.* be 3; therefore, *.ag.* is 4 and *.dg.* is 2, and from the multiplication of *.ag.* with *.ab.* results *.ez.*, and from *.ab.* times *.dg.* results *.zi.*; therefore, as *.ag.* is to *.dg.*, so is *.ez.* to *.zi.*. And because *.ab.* is 1, the numbers *.ez.* and *.zi.* equal the numbers *.ag.* and *.dg.*. Therefore, *.ez.* is 4 and *.zi.* is 2. Again, the multiplication of *.bg.* by *.ag.* yields *.km.*, and the multiplication of *.bg.* by *.gd.* yields *.kl.*; therefore, *.km.* is 12 and *.kl.* is 6. Therefore, as *.ez.* is to *.zi.*, that is as *.ag.* to *.dg.*, so is *.km.* to *.kl.*. And from the multiplication of *.ez.* and *.kl.* results *.op.*, and from the multiplication of *.zi.* and *.km.* results *.pq.*. For, indeed, *.op.* and *.pq.* are equal. Each of them is 24. For *.op.* is the sum of four consecutive odd numbers which are about *.kl.*, 6; these are about *.kl.* because of *.ez.*, which is 4; 4 added to *.kl.* is *.kn.*; therefore, *.kn.* is 10. Similarly, 4 subtracted from *.kl.* leaves *.kh.*; therefore, *.kh.* is 2. Again, the addition of the number *.zi.*, namely 2, to *.km.* makes *.kc.*; therefore, *.kc.* is 14 and *.nc.* is 4. There are two odd numbers, namely, 11 and 13 [between 10 and 14]; and they are placed about the number *.km.*, namely about 12, and they sum to the number *.pq.*. Therefore, the first square, namely *.ro.*, is 1, of which the root, namely 1, is half the number *.kh.*. Moreover, the second square, namely *.rp.*, is 25, of which the root, that is 5, is half the number *.kn.*. In fact, the third square, namely *.rq.*, is 49, of which the root, which is 7, is half the number

.*kc*.. And note that because 24 is the first congruous number, the smallest whole number which is a multiple of twenty-four, it results from the smallest number pair adding to make an even number, namely one and 3.

Alternatively, let the sum of .*ab*. and .*bg*. be an odd number. Subtract .*ab*. from .*bg*. and let the

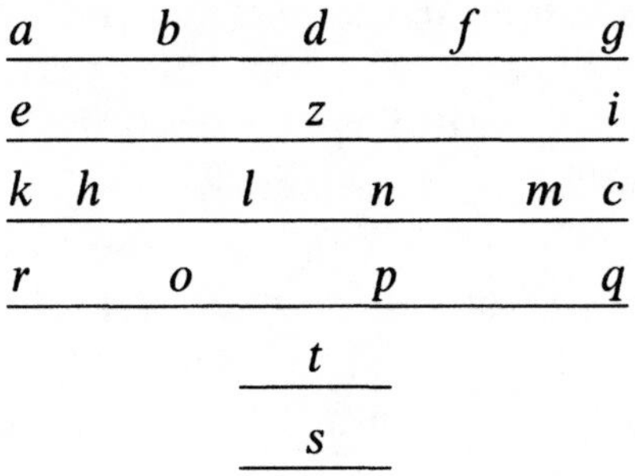

a b d f g

e z i

k h l n m c

r o p q

t

s

difference be the number .*gd*.. And let the ratio of the number .*gb*. to .*ba*. be smaller than the ratio of .*ag*. to .*gd*.. I shall find again two sets of consecutive odd numbers, their numbers in the ratio of .*gb*. to .*ba*.; from one quantity will be created again a congruous number; let, therefore, the number .*t*. be twice the number .*bg*., and the number .*s*. twice the number .*ba*.. The sum, in fact, of the numbers .*t*. and .*s*. is even, and .*t*. is to .*s*. as .*bg*. is to .*ba*.; and from the multiplications of .*gd*. by the numbers .*t*. and .*s*. result .*ez*. and .*zi*.. Therefore, as .*t*. is to .*s*., so is .*gb*. to .*ba*., as well as .*ez*. to .*zi*.. And again from the multiplications of the number .*ag*. by numbers .*t*. and .*s*. result .*km*. and .*kl*.. Therefore, as .*t*. is to .*s*., so is .*ez*. to .*zi*., as well as .*km*. to .*kl*., and because the numbers .*t*. and .*s*. are even, the numbers .*km*. and .*kl*. are even. And what is said in the premises about the numbers .*kh*., .*hl*., .*ln*., .*nm*. and .*mc*. is shown, and also about .*ro*., .*op*., .*pq*.. Therefore, .*op*. and .*pq*. are congruous numbers, etc. We show this moreover with numbers.

Let *.bg.* be 2 and *.ab.* be 1. Then *.ag.* will be 3 and *.gd.* will be 1, and *.t.* 4 and *.s.* 2 and *.ez.* 4 and *.zi.* 2 and *.km.* 12 and *.kl.* 6 and *.hn.* 8 and *.nc.* 4 and *.kh.* 2. The square *.ro.* will be 1, the congruous numbers *.op.* and *.pq.* 24. Therefore, the square *.rp.* is 25 and the square *.rq.* 49. This is what had to be shown.

And because the numbers *.bg.* and *.ba.*, namely one and two, are the smallest of the numbers; and the addition of them yields an odd number; and from them we had 24 as a congruous number; and we had before 3 and 1, which are the smallest numbers that can make the sum even. For that reason 24 is manifestly the smallest and first congruous number which falls among three squares of whole numbers. But for fractions can be found smaller numbers, as we shall demonstrate later.

But if the ratio of *.gb.* to *.ba.* is greater than the ratio of *.ag.* to *.gd.*, then the number of odd numbers in the first congruous number is to the number in the second as *.ag.* is to *.gd.*. Let again, *.ag.* be odd. To demonstrate this clearly, let *.gb.* be 5 and *.ba.* be 2; therefore, *.ag.* will be 7 and *.dg.* will be 3. Moreover, twice *.gb.* is, in fact, 10 and twice *.ba.* is 4; and the multiplying of 4 with *.ag.* and *.gd.* results in 28 for the number of odd numbers in the first congruous number and 12 for the number in the second. And multiplying each of the numbers *.ag.* and *.dg.* by twice *.bg.*, you will have 70 for the number about which are located 12 odd numbers making the second congruous number, and 30, about which there will be 28 odd numbers making the first. Therefore, subtract 28 from 30 leaving 2, of which the half, namely one, is the root of the first square; and adding 28 to 30, there will be 58, of which the half, namely 29, is the root of the second square. Again, add 12 to 70 making 82, of which the half, namely 41, is the root of the third square; and from the multiplication of 28 by 30, as well as 12 by 70, is had 840 for the congruous number.

Again, 840 results from two other given numbers,

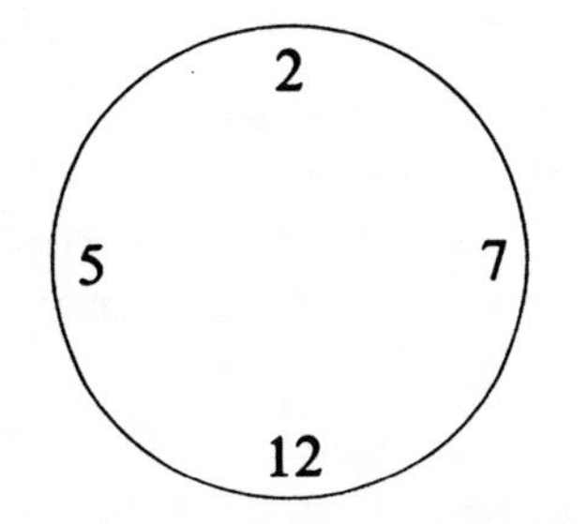

namely from seven and five, because if the triple product made from them and their sum is multiplied by two, namely 7 minus 5, there will result 840. In fact, the first collection of odd numbers which make this number, is in quantity two times seven. Therefore, the number of odd numbers is 14, and they are located about 60, which results from five times 12. Moreover, the number of the second collection results from two times five, and they are located about 84, which results from the multiplication of seven and 12. Then ten times 84, as sixty times 14, equals the triple product made by 5, 7, 12, multiplied by two; thus, a multiple of twenty-four and a congruous number as is shown above.

The first congruous number that can be found with integral squares is 24, and from 24 are generated all congruous numbers. Indeed, how many times 24 shall be multiplied by a square number, as many congruous numbers will be produced, and among the three squares which are congruent there will be the smallest square which multiplied 24. Moreover, the middle square will be the number which will be produced by the same square multiplied by 25. The third square, in fact, will be the number which will result from the first square multiplied by 49, of which the root will be the number which is the product of the root of the first square by seven, which is the root of 49; and twice this number will be the number of odd numbers making the congruous number. Similarly, a congruous number will result if 24 will be multiplied by a sum of squares which will be made of a sum of increasing numbers, both odd and even beginning with the unity, or by odd numbers only, or by those which increase by threes, or by evens, fours, or by some remaining number for which the sum of the squares can be found as we explained before; and there will be a ratio of odd numbers making the second congruous number to the odds making the first as the root of the last square is to the root of the square which follows in the assumed collection. For example, the sum of the squares of three odd numbers, namely one and nine and twenty-five, is 35, which multiplied by 24 yields 840, which is a congruous number, a congruous number which results also from taking five and seven. Therefore the ratio of the first number of odd numbers to the second number is as 7 to 5, as is shown above.

Comments on Proposition 14

Leonardo's argument with line segments is very involved and somewhat difficult to follow. For that reason we will present the argument in modern notation first. The problem or proposition is to find three squares and a number c, which Leonardo calls the congruous number, so that

$$x^2 + c = y^2, y^2 + c = z^2.$$

Leonardo sees the solution to this problem in sums of odd numbers since each square is the sum of odd numbers.

$$[1 + 3 + \cdots + (2x - 1)] + c = [1 + 3 + \cdots + (2y - 1)].$$

$$[1 + 3 + \cdots + (2y - 1)] + c = [1 + 3 + \cdots + (2z - 1)].$$

If a number c can be found, it is simultaneously equal to two different sums of odd numbers.

$$c = (2x + 1) + (2x + 3) + \cdots + (2y - 1).$$

$$c = (2y + 1) + (2y + 3) + \cdots + (2z - 1).$$

The odd numbers in the first representation of c are smaller than those in the second, and therefore there must be more of them. In the first representation, there are $y - x$ odd numbers centered at $y + x$ giving a sum

$$(y - x)(y + x) = y^2 - x^2.$$

In the second representation, there are $z - y$ odd numbers centered at $z + y$, giving a sum

$$(z - y)(z + y) = z^2 - y^2.$$

The ratio of the number of odd numbers in the first representation to that of the number of odd numbers in the second representation is $(y - x)/(z - y)$. For example,

$$1^2 + 24 = 5^2. \qquad 24 = 3 + 5 + 7 + 9.$$

$$5^2 + 24 = 7^2. \qquad 24 = 11 + 13.$$

The ratio is $(5 - 1)/(7 - 5) = 4/2$.

We begin then with two given numbers, whole numbers m, n, $n > m > 0$. In the first case, we consider m and n to have the same parity, both odd or both even, and $n/m < (n + m)/(n - m)$. This implies $m(n + m) - n(n - m) > 0$. About the central position, $m(n + m)$, an even number, place $n(n - m)$ consecutive odd numbers; they will sum to $m(n + m)n(n - m)$. About a second central position, $n(n + m)$, also an even number, place $m(n - m)$ consecutive odd numbers; they will sum to $m(n - m)n(n + m)$ also. The ratio of the number of odd numbers in the first collection to the number of odd numbers in the second is

$$[n(n - m)]/[m(n - m)] = n/m.$$

The odd numbers in the first set will be

$$m(n + m) - n(n - m) + 1, \ldots, m(n + m) - 1, m(n + m) + 1,$$

$$n(n + m) + 3, \ldots, n(n + m) + n(n - m) - 1.$$

This collection can be rewritten as

$$2[m(n + m) - n(n - m)]/2 - 1, \ldots, 2[m(n + m)]/2$$

$$- 1, 2[n(n + m) + 2]/2 - 1,$$

$$2[m(n + m) + 4]/2 - 1, \ldots, 2[m(n + m) + n(n - m)]/2 - 1.$$

The odd numbers in the second set will be

$$2[n(n + m) - m(n - m)]/2 + 1, 2[n(n + m) - m(n - m) + 2]/2 + 1,$$
$$\ldots, 2[n(n + m) + m(n - m)]/2 - 1.$$

The odd numbers in the first collection total $m(n + m)n(n - m)$. The odd numbers in the second collection total $n(n + m)m(n - m)$. The odd numbers from the unity 1 up to $2[(n + m) - n(n - m)]/2 - 1$ total

$$\{[m(n + m) - n(n - m)]/2\}^2.$$

The odd numbers from the unity 1 up to $2[m(n + m) + n(n - m)]/2 - 1$ total

$$\{[m(n + m) + n(n - m)]/2\}^2.$$

The odd numbers from the unity 1 up to $2[n(n + m) + m(n - m)]/2 - 1$ total

$$\{[n(n + m) + m(n - m)]/2\}^2.$$

These yield two equations when the odd numbers are summed.

$$\{[m(n + m) - n(n - m)]/2\}^2 + mn(n - m)(n + m)$$
$$= \{[m(n + m) + n(n - m)]/2\}^2.$$
$$\{[n(n + m) - m(n - m)]/2\}^2 + mn(n - m)(n + m)$$
$$= \{(n(n + m) + m(n - m)]/2\}^2.$$

Note that

$$m(n + m) - n(n - m) = m^2 + 2mn - n^2$$

is a positive even number because $n/m < (n + m)/(n - m)$. Note also

$$m(n + m) + n(n - m) = m^2 + n^2 = n(n + m) - m(n - m).$$

This shows that the two collections of consecutive odd numbers are adjacent; there are two odd numbers separating the two collections of odd numbers. Finally, note

$$n(n + m) + m(n - m) = n^2 + 2mn - m^2$$

is always an even positive number.

Still considering the case when m and n both have the same parity and $n > m > 0$, we now look at what happens when

$n/m > (n + m)/(n - m)$; that is when $n(n - m) - m(n + m) > 0$. Again, there are two sets of consecutive odd numbers. There are $m(n + m)$ consecutive odd numbers centered about the number $n(n - m)$:

$$2\{[n(n-m)-m(n+m)]/2\} + 1, \ldots, 2\{[n(n - m) + m(n + m)]/2\} - 1.$$

And there are $m(n - m)$ consecutive odd numbers centered about the number $n(n + m)$:

$$2\{[n(n + m) - m(n - m)]/2\} + 1, \ldots,$$
$$2\{[n(n + m) + m(n - m)]/2\} - 1.$$

The ratio of the number of odd numbers in the first collection to the number of odd numbers in the second collection is

$$m(n + m)/m(n - m) = (n + m)/(n - m).$$

The sum of all the odd numbers in the first collection is

$$m(n + m)n(n - m)$$

and the sum of all the odd numbers in the second collection is $m(n - m)n(n + m)$. These yield two equations that solve the original problem.

$$\{[n(n - m) - m(n + m)]/2\}^2 + nm(n - m)(n + m)$$
$$= \{[n(n - m) + m(n + m)]/2\}^2.$$
$$\{[n(n + m) - m(n - m)]/2\}^2 + nm(n - m)(n + m)$$
$$= \{[n(n + m) + m(n - m)]/2\}^2.$$

Note here that

$$n(n - m) - m(n + m) = n^2 - 2mn - m^2 > 0$$

because $n/m > (n + m)/(n - m)$. Furthermore,

$$n(n - m) + m(n + m) = n(n + m) - m(n - m).$$

Finally, $n(n + m) + m(n - m)$ is still positive.

We now consider the case of two positive whole numbers m and n with $n > m > 0$, m and n relatively prime, but one number is odd and the other is even; this means m and n are of opposite parity. As before, we first consider $n/m < (n + m)/(n - m)$. This alters the previous argument in that $(n + m)$ and $(n - m)$ are both odd numbers. The following key numbers are then all odd: $m(n + m) - n(n - m)$, $m(n + m) + n(n - m)$ or $n(n + m) - m(n - m)$, $n(n + m) + m(n - m)$.

The consecutive odd numbers lying in the first collection are

$$2[m(n + m) - n(n - m)] + 1, \ldots, 2[m(n + m) + n(n - m)] - 1.$$

There are $2n(n - m)$ of these consecutive odd numbers centered about the number $2m(n + m)$. Their sum will be $2n(n - m)2m(n + m)$. The consecutive odd numbers lying in the second collection are

$$2[n(n + m) - m(n - m)] + 1, \ldots, 2[n(n + m) + m(n - m)] - 1.$$

There are $2m(n - m)$ odd numbers in the second collection centered about the number $2n(n + m)$. Their sum will be $2m(n - m)2n(n + m)$. The ratio of numbers in the first collection to the number of numbers in the second collection is $2n(n - m)/2m(n - m) = n/m$. The equations which solve the original problem are

$$\begin{aligned}
&[m(n + m) - n(n - m)]^2 + 4nm(n - m)(n + m) \\
&\quad = [m(n + m) + n(n - m)]^2; \\
&[n(n + m) - m(n - m)]^2 + 4nm(n - m)(n + m) \\
&\quad = [n(n + m) + m(n - m)]^2.
\end{aligned}$$

Now in case $n/m > (n + m)/(n - m)$, the first number must be chosen to be $n(n - m) - m(n + m)$ in order to be positive. The solutions x, y and z are therefore

$$\begin{aligned}
&n(n - m) - m(n + m), \\
&n(n - m) + m(n + m) \quad \text{or} \quad n(n + m) - m(n - m), \\
&n(n + m) + m(n - m).
\end{aligned}$$

The first collection of odd numbers runs from $2[n(n - m) - m(n + m)] + 1$ to $2[n(n - m) + m(n + m)] - 1$. There are $2m(n + m)$ of odd numbers centered about the number $2n(n - m)$. The second collection of consecutive odd numbers is from $2[m(n + m) - n(n - m)] + 1$ to $2[m(n + m) + n(n - m)] - 1$. There are in the second collection

$2m(n-m)$ odd numbers centered at $2n(n+m)$. The ratio of the number of odd numbers in the first collection to the number of odd numbers in the second collection is

$$2m(n+m)/2m(n-m) = (n+m)/(n-m).$$

The two equations containing the solution are

$$[n(n-m) - m(n+m)]^2 + 4nm(n-m)(n+m)$$
$$= [n(n-m) + m(n+m)]^2;$$
$$[n(n+m) - m(n-m)]^2 + 4nm(n-m)(n+m)$$
$$= [n(n+m) + m(n-m)]^2.$$

Having seen Leonardo's proofs in modern notation, we now discuss his actual proofs as he gives them with line segments. The first proof given is that when m and n have the same parity; that is, $n+m$ is an even number, and when $n/m < (n+m)/(n-m)$. The notation in Leonardo's presentation is as follows:

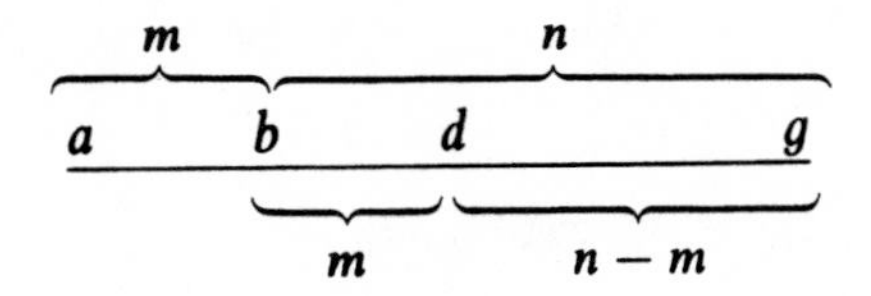

$m = ab. \qquad n = bg. \qquad n+m = ag. \qquad n-m = dg.$

$n/m = (bg)/(ab) < (ag)/(dg) = (n+m)/(n-m).$

$n(n-m) = (bg)(dg) = (ez).$

$m(n-m) = (dg)(ab) = (zi).$

$n(n+m) = (ag)(gb) = (km).$

$m(n+m) = (ag)(ab) = (kl).$

$(bg)/(ab) = (ez)/(zi)$

$(bg)/(ab) = (km)/(kl).$

$n(n-m)m(n+m) = (ez)(kl) = (op).$

$m(n-m)n(n+m) = (zi)(km) = (pq).$

About $(kl) = m(n + m)$ there are placed symmetrically $(ez) = n(n - m)$ consecutive odd numbers which sum to

$$(ez)(kl) = (op) = n(n - m)m(n + m).$$

About $(km) = n(n + m)$ there are placed symmetrically $(zi) = m(n - m)$ consecutive odd numbers, which sum to

$$(zi)(km) = (pq) = m(n - m)n(n + m).$$

$$(kh) = (kl) - (lh) = (kl) - (ez) = m(n + m) - n(n - m).$$

$$(kn) = (kl) + (ln) = (kl) + (ez) = m(n + m) + n(n - m)$$

$$= n(n + m) - m(n - m).$$

$$(kc) = (km) + (mc) = (km) + (zi) = n(n + m) + m(n - m).$$

$$[(kh)/2]^2 + op = [(kn)/2]^2. \qquad (ro) + (op) = (rp).$$

$$[(kn)/2]^2 + pq = [(kc)/2]^2. \qquad (rp) + (pq) = (rq).$$

For the case just presented, $n/m < (n + m)/(n - m)$, $(n + m)$ an even number, Leonardo works out this example.

$$m = 3. \qquad n = 5. \qquad n + m = 8. \qquad n - m = 2.$$

$$n/m = 5/3 < 8/2 = (n + m)/(n - m).$$

$$n(n - m) = 10. \qquad m(n - m) = 6.$$

$$n(n + m) = 40. \qquad m(n + m) = 24.$$

$$n(n - m)m(n + m) = 240.$$

About 24 are placed symmetrically 10 consecutive odd numbers 15, 17, 19, 21, 23, 25, 27, 29, 31, 33 which sum to 240. About 40 are placed symmetrically 6 consecutive odd numbers 35, 37, 39, 41, 43, 45 which also sum to 240.

$$(1/2)[m(n + m) - n(n - m)] = (1/2)(14) = 7.$$

$$(1/2)[m(n + m) + n(n - m)] = (1/2)(34) = 17.$$

$$(1/2)[n(n + m) + m(n - m)] = (1/2)(46) = 23.$$

$$7^2 + 240 = 17^2. \qquad 17^2 + 240 = 23^2.$$

The case, $n/m > (n + m)/(n - m)$, $(n + m)$ an even number, is not proved by Leonardo but demonstrated by this example.

$m = 1$. $\quad n = 3$. $\quad n + m = 4$. $\quad n - m = 2$.

$n(n - m) = 6$. $\quad m(n - m) = 2$.

$n(n + m) = 12$. $\quad m(n + m) = 4$.

$n(n - m)m(n + m) = 24 \quad n/m = 3/1 > 4/2 = (n + m)/(n - m)$.

About 6 are placed symmetrically 4 consecutive odd numbers 3, 5, 7, 9 which total 24. About 12 are placed symmetrically 2 consecutive odd numbers, 11, 13, which total 24

$$(1/2)[n(n - m) - m(n + m)] = (1/2)[2] = 1.$$

$$(1/2)[n(n - m) + m(n + m)] = (1/2)[10] = 5.$$

$$(1/2)[n(n + m) + m(n - m)] = (1/2)[14] = 7.$$

$$1^2 + 24 = 5^2. \qquad 5^2 + 24 = 7^2.$$

Leonardo next briefly summarizes the proof for the case

$$n/m < (n + m)/(n - m)$$

and $n + m$ an odd number.

$m = ab$. $\quad n = bg$. $\quad n + m = ag$. $\quad n - m = dg$.

$n/m = (bg)/(ab) < (ag)/(dg) = (n + m)/(n - m)$.

$t = 2(bg)$. $\quad s = 2(ab)$. $\quad t + s$ is even. $\quad t/s = n/m$.

$2n(n - m) = 2(t)(gd) = (ez)$. $\quad 2m(n - m) = 2(s)(gd) = (zi)$.

$2n(n + m) = 2(t)(ag) = (km)$. $\quad 2m(n + m) = 2(s)(ag) = (kl)$.

$(bg)/(ab) = (ez)/(zi)$. $\quad (bg)/(ab) = (km)/(kl)$.

$2n(n - m)2m(n + m) = (ez)(kl) = (op)$.

$2m(n - m)2n(n + m) = (zi)(km) = (pq)$.

About $(kl) = 2m(n + m)$ there are placed symmetrically $(ez) = 2n(n - m)$ consecutive odd numbers which sum to $2n(n - m)2m(n + m) = (ez)(kl) = (op)$. About $(km) = 2n(n + m)$ there are placed symmetrically $(zi) = 2m(n - m)$ consecutive odd numbers which sum to $(zi)(km) = (pq) = 2m(n - m)2n(n + m)$.

$$(kh) = (1/2)[2m(n + m) - 2n(n - m)].$$

$$(kn) = (1/2)[2m(n + m) + 2n(n - m)]$$

$$= (1/2)[2n(n + m) - 2m(n - m)].$$

$$(kc) = (1/2)[2n(n + m) + 2m(n - m)].$$

$$[(kh)/2]^2 + op = [(kn)/2]^2. \qquad (ro) + (op) = (rp).$$

$$[(kn)/2]^2 + pq = [(kc)/2]^2. \qquad (rp) + (pq) = (rq).$$

Leonardo next gives the following example of this just proven case.

$$m = 1. \qquad n = 2. \qquad n + m = 3. \qquad n - m = 1.$$

$$n/m = 2/1 < 3/1 = (n + m)/(n - m).$$

$$2n(n - m) = 4. \qquad 2m(n - m) = 2$$

$$2n(n + m) = 12. \qquad 2m(n + m) = 6.$$

$$2n(n - m)2m(n + m) = 24.$$

About 6 are placed 4 consecutive odd numbers, 3, 5, 7, 9, which sum to 24. About 12 are placed 2 consecutive odd numbers, 11, 13, which sum to 24.

$$(1/2)[2m(n + m) - 2n(n - m)] = (1/2)(2) = 1.$$

$$(1/2)[2m(n + m) + 2n(n - m)] = (1/2)(10) = 5.$$

$$(1/2)[2n(n + m) + 2m(n - m)] = (1/2)(14) = 7.$$

$$1^2 + 24 = 5^2 \qquad 5^2 + 24 = 7^2.$$

The case for $n/m > (n + m)/(n - m)$, $(n + m)$ an odd number, Leonardo does not prove but treats by giving this example.

$ab = m = 2.$ $bg = n = 5.$ $ag = n + m = 7.$ $dg = n - m = 3.$

$n/m = 5/2 > 7/3 = (n + m)/(n - m).$

$2n(n - m) = 30.$ $2m(n - m) = 12.$

$2n(n + m) = 70.$ $2m(n + m) = 28.$

$2n(n - m)2m(n + m) = 840.$

About 30 are placed 28 consecutive odd numbers, 3, 5, . . . , 55, 57, which sum to 840. About 70 are placed 12 consecutive odd numbers, 59, 61, . . . , 79, 81, which sum to 840.

$$(1/2)[2n(n - m) - 2m(n + m)] = (1/2)(2) = 1.$$

$$(1/2)[2n(n - m) + 2m(n + m)] = (1/2)(58) = 29.$$

$$(1/2)[2n(n + m) + 2m(n - m)] = (1/2)(82) = 41.$$

$$1^2 + 840 = 29^2. \qquad 29^2 + 840 = 41^2.$$

A method for an algebraic solution for this problem in proposition 14 can be found in the work of Diophantus. In Book III, problem 19, and in Book V, problem 7, of the *Arithmetica* [H1, pp. 166, 205], Diophantus states the identity,

$$(\text{hypoteneuse})^2 \pm \text{twice product of perpendiculars} = \text{a square}.$$

This equation of Diophantus provides the solution to the problem in this proposition and proposition 17. Using the sides of a right triangle as

$$m^2 - n^2, 2nm, m^2 + n^2,$$

the identity is expressed symbolically as

$$(m^2 + n^2)^2 \pm 4nm(n^2 - m^2) = (n^2 - m^2 \pm 2nm)^2.$$

The problem of proposition 17, $x^2 \pm 5 =$ a square, a consequence of proposition 14, can be solved using Diophantus' equation as follows. Set $n = 5$ and $m = 4$.

$$(25 + 16)^2 \pm (4)(5)(4)(9) = (25 - 16 \pm 40)^2.$$

$$41^2 \pm (144)(5) = (9 \pm 40)^2.$$

$$41^2 + 5(12) = 49^2. \qquad 41^2 - 5(12) = 31^2.$$

$$(41/12)^2 + 5 = (49/12)^2. \qquad (41/12)^2 - 5 = (31/12)^2.$$

It is inconceivable that Diophantus would not have made short work of this problem. The numbers obtained using Diophantus' methods are the same as those obtained by Leonardo. Diophantus did not work the problem, however, and Leonardo's methods are different. What is different about Leonardo's work is the use and interplay of the sequences of odd numbers to obtain the proper relation between the squares. The problem shows up also in Arabic works following Diophantus. It was analyzed, for example, by the mathematician al-Khazin [A, p. 49]. Also, Alkarkhi's treatment of the problem as well as some of Leonardo's work is discussed by Franz Woepcke [Wo, p. 143]. A complete history of the problem is given by L. E. Dickson [D, p. 459ff.]. Oystein Ore also provides a rewarding discussion of Leonardo's work on this problem [O, pp. 185–193].

Proposition 15

If some congruous number and its congruent squares are multiplied by another square, the number made by the product of the congruous number and the square will be a congruous number; the remaining squares will be congruent with this congruous number.

Let .*ab*. be a square, and .*bg*. be a congruous number, and .*gd*. be equal to the number .*bg*.. Therefore, the numbers .*ag*. and .*ad*. will be squares. And let, moreover, .*e*. be a square number.

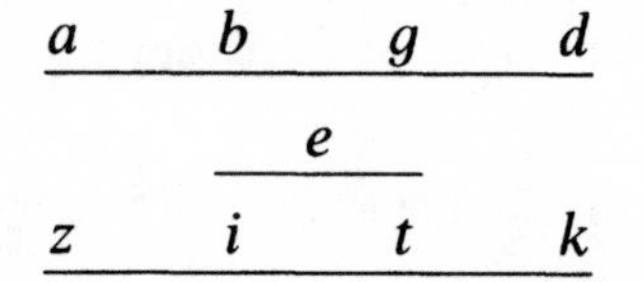

I say that the product of .*e*. with the congruous number .*bg*. will be a congruous number; and the numbers made by multiplying *.e.* with *.ab.*, *.ag.*, *.ad.* will be squares made congruent by the congruous number *.e.* times *.bg.*. In fact, from the multiplication of *.e.* and *.ab.* results *.zi.*; and from *.e.* times *.ag.* results *.zt.*; and from *.e.* times *.ad.* results *.zk.*. And because *.e.* and *.ag.* are squares so also is the product *.zt.* square. Therefore, *.zt.* is equal to the sum of two numbers, the square *.e.* times the square *.ab.*, that is *.zi.*, and *.e.* times the congruous number *.bg.*. But the number *.zi.* is made from the multiplication of square *.e.* and square *.ab.*. Therefore, the difference *.it.* is produced from *.e.* times the congruous number *.bg.*. And because the numbers *.e.* and *.ab.* are squares, their product is a square; the number *.zi.* is therefore a square. Again, because the number *.zk.* is the product of the square *.e.* and the square, *.ad.*, so is the number *.zk.* a square and is equal to the sum of two numbers, *.e.* times *.ag.* and *.e.* times *.gd.*. But *.e.* times *.ag.* is the square *.zt.*. Therefore, the remaining *.tk.* is *.e.* times *.gd.*. And because the number *.bg.* and the number *.gd.* are equal, *.e.* times *.bg.* and *.e.* times *.gd.* will be equal. Also, *.e.* times *.bg.* is *.it.*. Therefore, *.it.* is equal to the number *.tk.*. If now to the square number *.zt.* is added the number *.tk.*, the sum is the square *.zk.*; and if from the square *.zt.* is subtracted *.tk.*, that is, *.ti.*, there will remain the square *.zi.*; therefore, *.it.* is a congruous number, and the three squares *.zi.* and *.zt.* and *.zk.* are congruent. This had to be shown. Similarly, the same will be shown if some congruous number and its congruent squares are divided by some square number.

Comments on Proposition 15

If c is a congruous number for the squares of x, y, and z, then

$$x^2 + c = y^2; y^2 + c = z^2.$$

If we multiply both equations by a square number, for example, the square of u, then we have

$$(xu)^2 + cu^2 = (yu)^2; (yu)^2 + cu^2 = (zu)^2.$$

Thus cu^2 is a congruous number for (xu), (yu), and (zu). Proposition 15 is self-evident in modern notation.

In line segment notation (with $bg = gd$),

$$ab + bg = ag; ag + gd = ad.$$

Multiplying by the square number e gives

$$(ab)(e) + (bg)(e) = (ag)(e);$$

$$(ag)(e) + (gd)(e) = (ad)(e).$$

Leonardo rewrites these as

$$zi + it = zt; zt + tk = zk.$$

Proposition 16

I wish to find a congruous number which is a square multiple of five.

Let one of two given numbers be 5; let the other be a square number so that the sum of them is a square number, and so that the lesser subtracted from the greater leaves a square number. For that reason there will be then a square 4, which added to five makes 9, a square; and 4 subtracted from 5 leaves 1, which is also a square. I say these two given numbers yield a congruous number which will be a square multiple of five. There results certainly a congruous number from the product of one times twice five times twice four times nine; this is multiplying the double product one times ten by the double product eight times nine, which is 10 times 72. But the product of four times nine is a square number, for both are squares. Therefore, the product of 8 times 9 is twice a square, and the multiplication of this double square by twice five results in four times a square times five. But four times a square is a square number; therefore, four times a square times five is five times a square, and five times this square times 1, a square, is again five times a square. Therefore, the congruous number made from these will be a square multiple of five.

Comments on Proposition 16

The Latin manuscript for the statement of this theorem reads *congruum cuius quinta pars sit integra*, that is a congruous number which is an integral multiple of five. Although this is not literally incorrect, it is clear Leonardo means the stronger, square multiple of five. The Italian Renaissance manuscript of Master Bendetto published by Mr. Ettore Picutti [P1, p. 245] reads square as well as integral multiple of five.

Leonardo wishes to find a congruous number which is a square multiple of 5 and does so. He uses his result in the next proposition. He chooses n, m so that $nm(n - m)(n + m)$ or rather $4nm(n - m)(n + m)$ is a square multiple of 5. The choice is $n = 5, m = 4$, which makes $n + m$ an odd number and leads to $4nm(n - m)(n + m)$ equal to 720. 720 is 144 times 5, a square multiple of 5.

Proposition 17

Here is the question mentioned in the prologue of this book.

I wish to find a square number which increased or diminished by five yields a square number.

Take a congruous number, a square multiple of five; 720 will be one, of

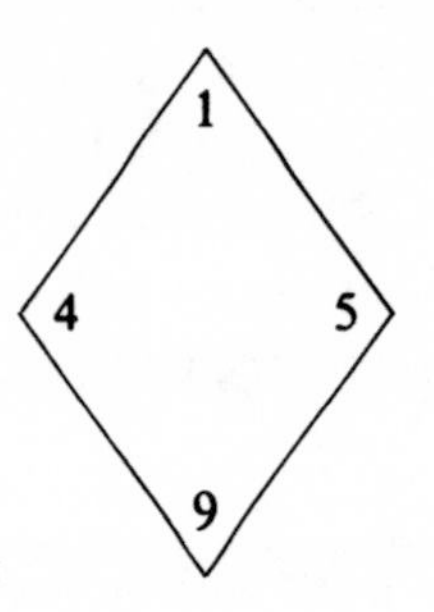

which the fifth part is 144, by which divide the same 720 and the congruent squares, of which the first is 961, the second is 1681, the third indeed is 2401. The root of the first square is 31, the second 41, third 49. There is for the first square 6 97/144, with root 2 7/12, which results from dividing 31 by the root of 144, which is 12, and there is for the second, which is the sought square, 11 97/144, with root 3 5/12, which results from dividing 41 by 12, and there is for the last square 16 97/144 with root 4 1/12.

Comments on Proposition 17

It is to be noted that the congruent square solutions sought by Leonardo in this theorem are not whole numbers but fractions. Since 5 is not a multiple of 24, and therefore not a congruous number, there can be no integral solutions to

$$x^2 + 5 = y^2 \quad \text{and} \quad y^2 + 5 = z^2.$$

Rational solutions, however, can be found by choosing n, m so that the congruous number is a square multiple of 5. The choice $n = 5$, $m = 4$ yields the following solution of the problem solved in proposition 14.

Obviously, proposition 15 and 16 prepare for this solution.

$$m = 4. \qquad n = 5. \qquad n + m = 9. \qquad n - m = 1.$$

$n + m$ is an odd number. $\quad n/m = 5/4 < 9/1 = (n + m)/(n - m)$.

$$2n(n - m) = 10. \qquad 2m(n - m) = 8.$$
$$2n(n + m) = 90. \qquad 2m(n + m) = 72.$$

$$(1/2)[2m(n + m) - 2n(n - m)] = (1/2)(62) = 31.$$
$$(1/2)[2m(n + m) + 2n(n - m)] = (1/2)(82) = 41.$$
$$(1/2)[2n(n + m) + 2m(n - m)] = (1/2)(98) = 49.$$

$$31^2 + 720 = 41^2. \qquad 41^2 + 720 = 49^2.$$

Division by 144, or the square of 12, yields this solution.

$$(1/12)^2 + 5 = (41/12)^2. \qquad (41/12)^2 + 5 = (49/12)^2.$$

In Master Benedetto's work [P1, p. 249] of ca. 1464, he presents a list of congruous numbers and congruent squares calculated by himself and presumably others. Leonardo's successors in the Tuscan school were interested in the question of which integers C can be congruous numbers and which cannot be, as are mathematicians today. Unable to find a general answer to this question, Leonardo's successors tabulated their results and the table of integral congruent squares and congruous numbers appears supplementarily to the translation of Leonardo's work. Luca Pacioli in his *Summa* discusses the problem along with his summary of *Liber quadratorum*. He also provides a table of then known results.

A slightly different question arises when one asks for which positive integers C can be found *rational* numbers X, Y, Z so that

$$Y^2 + C = Z^2 \quad \text{and} \quad Y^2 - C = X^2.$$

Leonardo has shown that 5 is such an integer in solving the question posed in proposition 14. Since 5 is not a multiple of 24 and the X, Y, Z found are not integers, 5 should not properly be called a congruous number. Terminology has varied, but modern researchers seem to use the term *congruent number* for such solutions. This should not, of course, be confused with congruent as used by C. F. Gauss, as in x is congruent to y modulo p, meaning $x - y$ is a multiple of p.

In his history *Number Theory* [We, p. 13], Mr. Weil points out that the problem can be reduced to one of finding Pythagorean triples, but there is no indication that Leonardo made this observation. Here is the reasoning.

$$Y^2 - X^2 = Z^2 - Y^2 = C. \qquad 2Y^2 = Z^2 + X^2.$$

$$Y^2 = (1/2)[Z^2 + X^2] = [(Z + X)/2]^2 + [(Z - X)/2]^2.$$

Setting $U = (Z + X)/2$ and $V = (Z - X)/2$, we have then

$$Y^2 = U^2 + V^2 \quad \text{with} \quad Y^2 - X^2 = 2UV = C.$$

$$X = U - V; Z = U + V.$$

U, V and Y are then Pythagorean triples and the number C is four times the area of the Pythagorean right triangle. From any Pythagorean triple U, V and Y one can obtain a solution to the original equations.

The general question of which positive integers can be congruent numbers has not been resolved even to this day. The question can be restated to one of elliptic curves. Mr. Weil further points out that the known solutions for Pythagorean triples

$$U = d2mn, V = d(m^2 - n^2), Y = d(m^2 + n^2), m > n > 0,$$

yield Pythagorean triangle area $(1/2)UV = d^2mn(m^2 - n^2)$, which gives

$$C = 4d^2mn(m^2 - n^2).$$

If we set $t = Cm/n$, we get

$$t^3 - C^2t = (Cm/n)^3 - C^2(Cm/n) = C^3m(m^2 - n^2)/n^3 = C^4/4d^2n^4$$
$$= (C^2/2dn^2)^2 = w^2,$$

with $w = C^2/2dn^2$. We have therefore (t, w) as a rational point on the curve $t^3 - C^2t = w^2$. Conversely, if (t, w) is any rational point on the elliptic curve, a reverse argument produces a Pythagorean triple which will have triangle area a d^2 multiple of C. Therefore, the congruent numbers are those which lead to rational points (t, w), $w \neq 0$, on the elliptic curve $w^2 = t^3 - C^2t$.

Mr. John Coates [Bourbaki Seminar Lecture 635, November 1984] has summarized the present state of research for general results on congruent numbers. It is presently conjectured on the basis of strong evidence, but not yet proven, that C will be a congruent number if and only if C is congruent (in the Gaussian sense) to 5, 6 or 7 modulo 8. On the negative side, it is proven that if C is a prime congruent to 3 modulo 8, then C will not be a congruent number. Likewise, if C is double a prime congruent to 5 modulo 8, then C will not be a congruent number. Finally, if C is a prime congruent to 9 modulo 16 and 2 is fourth power modulo C, or C is congruent to 1 modulo 16 and 2 is not a fourth power modulo C, then C is not a congruent number. I am indebted to Mr. André Weil for this communication.

Proposition 18

If any two numbers have an even sum, then the ratio of their sum to their difference, the larger less the smaller, will not be the same as the ratio of the larger number to the smaller.

Let any two numbers .ab. and .bg. be given. Let the greater .gb. exceed the lesser .ba. by .gd.. Let the number .ag. be even. I say that .gb. to .ba. will not be as .ag. to .gd.. Suppose it were possible that .ag. be to .gd. as .gb. is to .ba.. Then the product of .bg. and .gd. will be equal to the product of .ag. and .ba.. Let, therefore, the number .ze. be the product of .bg. and .gd., and the number .kl. be the product of .ag. and .ba.. Therefore, the number .ze. is equal to the number .kl.. However, from the multiplication of .ze. and .kl. results .op.. Therefore, the number .op. is equal to the

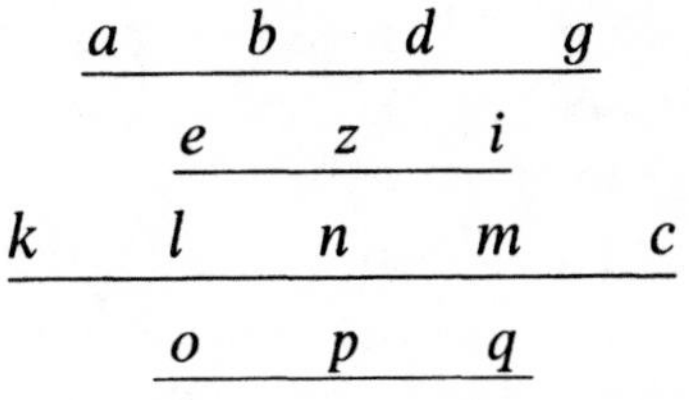

a b d g

e z i

k l n m c

o p q

number which is the product of *.ba.*, *.gd.*, *.ag.* and *.bg.*. Let, therefore, the number *.zi.* be *.ab.* times *.gd.*, and *.km.* be *.ag.* times *.bg.*; *.km.* is greater than *.kl.*, for the number *.gb.* is greater than *.ba.*. Therefore, *.op.* is equal to *.zi.* times *.km.*. For that reason, the number *.pq.* is *.zi.* times *.km.*; therefore, the number *.op.* is equal to the number *.pq.*. But *.op.* is a square because it is the product of two equal numbers, *.ez.* and *.kl.*. Therefore, *.pq.* is a square. It was shown above in finding the first congruous number that the number *.lm.* is equal to the number *.ei.*. Moreover, the number *.ei.* is greater than *.ez.*; therefore, *.lm.* is greater than *.ez.*. Therefore, the number *.ln.*, equal to the number *.ez.*, which is *.kl.*, is subtracted from the number *.lm.*. The difference *.nm.* is the number *.zi.*. Then the number *.mc.*, equal to the number *.mn.*, is added to *.km.*. And because *.kn.* is twice the number *.kl.*, *.kn.* will be even. In fact, the sum of the collection of odd numbers from one up to *.kn.* results in the square *.op.*. They are indeed located about the number *.kl.*, because *.kl.* is the root of the number *.op.*. Again, *.zi.* times *.km.* is the square *.pq.*. But as many unities as there are in the number *.zi.*, as many odd numbers there are between the number *.kn.* and *.kc.*. The number *.nc.* is, moreover, twice the number *.zi.*; therefore, the square *.pq.* is the sum of the odd numbers between *.kn.* and *.km.*. Therefore, the sum of the two squares *.op.* and *.pq.* is the sum of all the odd numbers from the unity up to the number *.kc.*. Therefore, the number *.oq.* is a square, and is twice the square *.op.*. The ratio, therefore, of the square *.oq.* to the square *.op.* is 2 to 1, that is, as a nonsquare number to a square number; that is contradictory. Therefore, the sum *.ag.* to the difference *.dg.* is not as *.gb.* is to *.ba.*. This is what had to be shown.

This same thing could have been demonstrated if the number *.ag.* were odd because the ratio of *.bg.* to *.ba.* is the same as twice *.gb.* to twice *.ba.*. Whence the number *.ez.*

could have been shown equal to the number .*kl*., etc. From this will be shown, in fact, that no square number can be a congruous number; because if it were possible, then the ratio of the sum of the two given numbers to the difference would be as the larger of them to the smaller.

Because of these reasons, there are many numbers which cannot be congruous numbers; but any number can be a congruous number if the division of some congruous number by this number results in a square number; or if this number is one of four given numbers and the remaining three are squares. If we take 9 and 16, which are squares, and the sum of them, namely, 25, is a square; and subtract 9 from 16 leaving 7; this 7 can be a congruous number. Multiply, in fact, twice 9 by twice 16 making a square number, namely, 576; which, is multiplied by 25, is again a square number; which, if multiplied by 7, makes a congruous number; therefore, it will be a square multiple of 7.

Comments on Proposition 18

It is to be proven that $n/m \neq (n + m)/(n - m)$ for any whole numbers n, m, $n > m > 0$. That this is true is seen in a few steps. Suppose $n/m = (n + m)/(n - m)$. Then $n(n - m) = m(n + m)$.

$$n^2 - nm = mn - m^2. \qquad m^2 + 2mn = n^2.$$

$$m^2 + 2mn + n^2 = 2n^2. \qquad (m + n)^2 = 2n^2.$$

But the ratio of the squares of whole numbers cannot be 2.

Leonardo obtains the same kind of contradiction but in a rather more complex fashion. Both (op) and (pq) represent the congruous number $n(n - m)m(n + m)$. If $n(n - m) = m(n + m)$, then

$$(op) = n(n - m)m(n + m) = n(n - m)n(n - m),$$

which is a square.

$$(op) = [n(n - m)]^2.$$

This square is of course the sum of odd numbers from 1 up to

$$\begin{aligned} 2[n(n-m)]-1 &= n(n-m)+n(n-m)-1 \\ &= m(n+m)+n(n-m)-1 \\ &= n(n+m)-m(n-m)-1. \end{aligned}$$

On the other hand, because (pq) is a congruous number, (pq) is the sum of odd numbers from $n(n+m)-m(n-m)+1$ up to $n(n+m)+m(n-m)-1$. If we add together (op) and (pq), producing (oq), it is the sum of odd numbers from 1 up to $n(n+m)+m(n-m)-1$. But this means

$$(oq)=\{[n(n+m)+m(n-m)]/2\}^2.$$

But $(oq)=(op)+(pq)=2(op)$. This means that $2(op)=(oq)$ and that both (op) and (oq) are squares of whole numbers. This is a contradiction. The case for $m+n$ odd is handled similarly.

It is apparent that if one is given 9, 16, 25, then

$$4nm(n-m)(n+m)=4(16)(9)(16-9)(16+9)=(120)(120)(7)$$

is a square multiple of 7 and therefore 7 is a congruous number for some rational squares.

Certainly, $n/m=(n+m)/(n-m)$ implies $nm(n+m)(n-m)$ or $4nm(n+m)(n-m)$ are square numbers, but did Leonardo really think that $4mn(n+m)(n-m)$ cannot be a square unless

$$n/m=(n+m)/(n-m),$$

as would be needed for the validity of his argument? His conclusion that no square can be a congruous number is equivalent to the fact that the area of a Pythagorean triangle cannot be a square, the proof of which was one of Fermat's greatest achievements [We, pp. 14, 77]. Undoubtedly, Leonardo was aware of the depth and difficulty of this question.

Proposition 19

I wish to find a square number for which the sum of it and its root is a square number and for which the difference of it and its root is similarly a square number.

Let a congruous number similarly be given with its three squares, numbers *.ab.*, *.ag.* and *.ad.*. Then the congruous number will be the numbers *.bg.* and *.gd.*. Each of the squares *.ab.*, *.ag.*, *.ad.* is divided by the congruous number *.bg.*, yielding numbers *.ez.*, *.ei.*, *.eh.*. And construct on *.ei.* a square *.ek.* and complete the area *.lh.*; and place *.z.* so that the segment *.zt.* is parallel to the segments *.ik.* and *.el.*; and then the number *.ez.* results from the division of the number *.ab.* by *.bg.*; and the number *.ei.* results from the division of the number *.ag.*

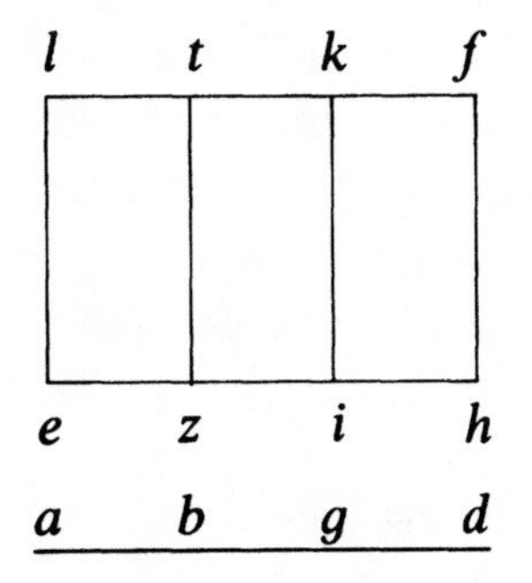

by the congruous number *.bg.*; in fact, the number *.zi.* results from the division of *.bg.* by itself. Therefore, *.zi.* is 1. Similarly, because *.ad.* is divided by the the congruous number *.gd.*, that is by *.bg.*, there results the number *.eh.*, and from the division of *.ag.* by *.gd.* results *.ei.*. Therefore, *.ih.* results from the division of *.gd.* by itself. Then *.ih.* is similarly 1; therefore, *.hi.* is equal to *.iz.*. And because on the segment *.ei.* the square *.ek.* is constructed, and *.hi.* is 1, the area of *.kh.* or *.kz.* is therefore the root of the area *.ek.*. Therefore, to the area of *.ek.* is added its root, namely the area of *.kh.*, and the result is the area of *.lh.*; and if from the square *.ek.* is subtracted the root, that is *.kz.*, there will remain the area *.zl.*. And because from the division of the numbers *.ab.* and *.ag.* and *.ad.* by some number, namely

.bg., there result numbers *.ez.*, *.ei.*, *.eh.*. Therefore, as *.ab.* is to *.ag.*, so is *.ez.* to *.ei.*; the numbers *.ab.* and *.ag.* are certainly square. Therefore, the ratio of the number *.ez.* to the number *.ei.* is as the ratio of a square number to a square number. Therefore, the multiplication of *.ez.* by *.ei.* results in a square number. But the segment *.el.* is equal to the segment *.zt.*, which is equal to the segment *.ik.*. The area *.ek.* is that of a square; therefore, the area *.et.* is a square number.

Similarly, because the ratio of *.ei.* to *.eh.*, that is *.le.* to *.eh.*, is as a square number to a square number, the product of *.eh.* and *.le.* will be a square, that is the area *.lh.*. A square number, *.ek.* in fact, is found with added root, that is *.kh.*, which makes square number *.lh.*; and if from the square number *.ek.* is subtracted its root, there will remain the square number *.et.*; this is what had to be done.

Comments on Proposition 19

The problem is to find numbers B, A, C so that

$$B^2 - B = A^2 \quad \text{and} \quad B^2 + B = C^2.$$

Suppose a solution x, y, z and a congruous number c are given according to proposition 14.

$$x^2 + c = y^2. \qquad y^2 + c = z^2.$$

$$y^2 - c = x^2. \qquad y^2 + c = z^2.$$

$$y^2/c - 1 = x^2/c. \qquad y^2/c + 1 = z^2/c.$$

$$(y^2/c)^2 - (y^2/c) = (xy/c)^2. \qquad (y^2/c)^2 + (y^2/c) = (yz/c)^2.$$

An example of this procedure is to take the solution

$$1 + 24 = 25. \qquad 25 + 24 = 49.$$

$$25/24 - 1 = 1/24. \qquad 25/24 + 1 = 49/24.$$

$$(25/24)^2 - (25/24) = (5/24)^2. \qquad (25/24)^2 + (25/24) = (35/24)^2.$$

In Leonardo's notation the argument is as follows. $(ab) + (bg) = (ag)$. $(ag) + (gd) = (ad)$. (ab), (ag), (ad) are squares and $(bg) = (gd)$ is the congruous number. $(ag) - (bg) = (ab)$. $(ag) + (gd) = (ad)$. Divide through by the congruous number (bg) or (gd). $(ag)/(bg) - 1 = (ab)/(bg)$. $(ag)/(bg) + 1 = (ad)/(bg)$. Set $(ab)/(bg) = (ez)$; $(ag)/(bg) = (ei)$; $(ad)/(bg) = (eh)$. $(ei) - 1 = (ez)$. $(ei) + 1 = (eh)$. Observing the geometrical diagram, we multiply by the side (el) to produce areas. $(el)(ei) - (el) = (el)(ez)$. $(el)(ei) + (el) = (el)(eh)$.

$$(el)^2 - (el) = (el)(ez). \qquad (el)^2 + (el) = (el)(eh).$$

Both $(el)(ez)$ and $(el)(eh)$ are squares. For example,

$$(el)(ez) = (ei)(ez) = [(ag)/(bg)][(ab)/(bg)] = [(ag)(ab)]/(bg)(bg).$$

Since both ag and ab are given squares, $(el)(ez)$ is also square.

Proposition 20

Similarly, a square number must be found which when twice its root is added or subtracted always makes a square number.

Given squares *.ab.*, *.ag.*, *.ad.*, are next divided by half the congruous number *.bg.* and there result the numbers *.ez.*, *.ei.*, *.eh.*; and the number *.zi.* will be 2, which will be equal to the number *.ih.*. Therefore, each of the areas *.kh.* and *.kz.* will be equal to twice the root of the square number *.ek.*; and similarly, the ratio *.ez.* to *.zt.* will be as the square *.ab.* to the square *.ag.*. Therefore, the product of the numbers *.ez.* and *.zt.*, that is the area *.et.* is a square; and because of this, the area, namely the number *.lh.*, is a square. Therefore, a square *.ek.* is found so that with the addition of two roots, namely *.kh.*, a square number *.lh.* results.

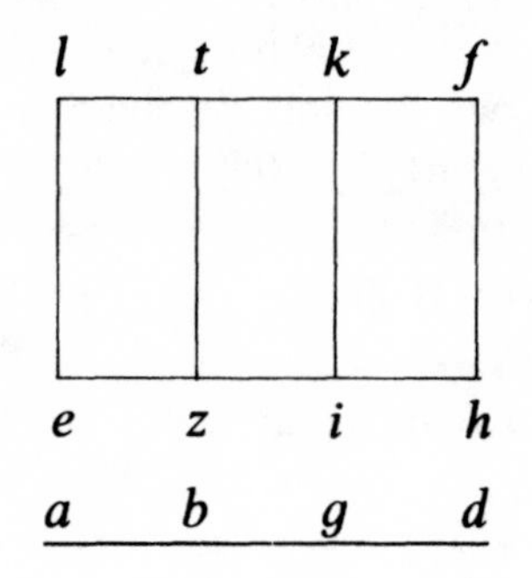

And when from the same square *.ek.* two roots are subtracted, namely *.kz.*, there remains the square number *.zl.*.

Again, the same result holds for three or more roots added or subtracted.

And to have the result in numbers, let the square *.ab.* be 1, and the square *.ag.* be 25. Also, let the square *.ad.* be 49. Therefore, the congruous number *.bg.* or *.gd.* will be 24; which is divided into 1 and 25 and 49 to yield *.ez.* 1/24; and *.ei.* will be 1 1/24. Also, the number *.eh.* will be 2 1/24. From the product of *.ei.* with itself, moreover, results the square *.ek.*, which is 625/576. If to it is added *.ki.* times *.ih.*, namely 1 1/24, there will result 1225/576, of which the root is 35/24, that is 1 11/24. Similarly, if one subtracts 1 1/24, that is 600/576, from 625/576, that is the number *.kz.* from the number *.ke.*, there will remain for the area *.et.*, 25/576, of which the root is 5/24. And if to some square it is proposed to add and subtract two roots, you will indeed double the numbers *.ez.*, *.ei.*, *.eh.*; there will result 1/12 and 2 1/12 and 4 1/12; which also will result if one divides 1, 25, 49, by half the congruous number, namely by 12; and thus the root of the sought square will be 2 1/12. And also the result holds for three or more roots added or subtracted.

Comments on Proposition 20

Proposition 20 is a variation of proposition 19. The equations to be solved are

$$B^2 - 2B = A^2 \quad \text{and} \quad B^2 + 2B = C^2.$$

The starting point is again proposition 14.

$y^2 - c = x^2.$	$y^2 + c = z^2.$
$y^2/c - 1 = x^2/c.$	$y^2/c + 1 = x^2/c.$
$2y^2/c - 2 = 2x^2/c.$	$2y^2/c + 2 = x^2/c.$
$[2y^2/c]^2 - 2[2y^2/c] = [2xy/c]^2.$	$[2y^2/c]^2 + 2[2y^2/c] = [2yz/c]^2.$

The solution is $B = 2y^2/c$, $A = 2xy/c$, $C = 2yz/c$.

The numerical example given is the following;

$1 + 24 = 25.$	$25 + 24 = 49.$
$25/24 - 1 = 1/24.$	$25/24 + 1 = 49/24.$
$25/12 - 1 = 1/12.$	$25/12 + 1 = 49/12.$
$(25/12)^2 - (25/12) = (5/12)^2.$	$(25/12)^2 + (25/12) = (35/12)^2.$

Proposition 21

For any three consecutive odd squares, the greatest square exceeds the middle square by eight more than the middle square exceeds the least square.

Let *.ab.*, *.bg.*, *.gd.* be three roots of three given consecutive odd squares; and let the root *.ab.* be less than *.bg.*, and

.bg. less than *.dg.*. I say that the square of the number *.gd.* exceeds the square of *.bg.* by eight more than the square of the number *.bg.* exceeds the square of the number *.ab.*. Because *.ab.*, *.bg.*, *.gd.* are roots of three consecutive odd squares, the roots are indeed odd and consecutive. Therefore, *.bg.* exceeds *.ab.* by two, and the number *.gd.* exceeds the number *.bg.* by the same number. Therefore, if two is subtracted from the number *.bg.* there remains the number *.ab.*. Likewise, if from *.gd.* is subtracted two, namely the number *.cd.* there will remain *.gc.* equal to the number *.bg.*. Again, if from the number *.cg.* is subtracted the number *.gf.*, equal to each of the numbers *.ab.* and *.be.*, there will remain two for the number *.fc.*. Therefore, the total *.fd.* is 4. And because the number *.ge.* is that by which the number *.bg.* exceeds the number *.ab.*; and *.eg.* is two; the square of the number *.bg.* exceeds the square of the number *.ab.* by twice the sum of the numbers *.ab.* and *.bg.*, that is *.ag.*. Similarly, the square of the number *.gd.* exceeds the square of the number *.bg.* by twice the sum of the number *.bg.* and *.gd.*. Therefore, the difference between the squares of the numbers *.gd.* and *.gb.* minus the difference between the squares of the numbers *.gb.* and *.ba.* is twice *.bg.* plus *.gd.* minus twice *.ab.* plus *.bg.*. From each is subtracted the double of the number *.bg.*;

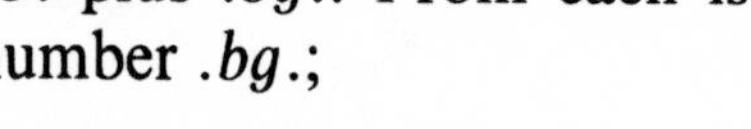

a b g f d

then twice the number *.gd.* exceeds twice the number *.ab.* by that which the difference of the squares of the numbers *.gd.* and *.gb.* exceeds the difference of the squares of the numbers *.gb.* and *.ba.*. But twice the number *.ab.* is twice the number *.gf.*; therefore, twice the number *.gd.* exceeds twice the number *.ab.*, that is twice the number *.gf.* by twice the number *.fd.*. But twice the number of *.fd.* is eight; for

.fd. is certainly four; therefore, the square of the number *.gd.* exceeds the square of the number *.bg.* by eight more than the square of the number *.bg.* exceed the square of the number *.ba.*. This had to be shown.

And because the second square is odd, namely 9, it exceeds by eight the first odd square, namely 1. It will be found that the third odd square, namely 25, exceeds the second odd square by twice eight. And thus it always will be found that the consecutive odd squares increase by a increasing sequence of eights. Likewise also for the even square numbers except for the second even square, namely 16, which exceeds by 12 the first even square, namely 4. Next, the even squares after 2 increase by eights to infinity. Even the second even square exceeds the first square by three fours; and the third exceeds the second by five fours; and the fourth exceeds the third by seven fours; and the fifth exceeds the fourth by nine fours; and thus always are added two fours by an increasing odd sequence of fours to infinity, according to this sequence. Even the first even square, namely 4, is made of one four. The second adds three fours to the first. The third adds five to the second. And thus it is found even squares increase by an increasing sequence of fours that are multiples of consecutive odd numbers. Similarly, I have found squares of numbers that increase by consecutive threes, which increase by consecutive odd multiples of nines.

For example, first the square of three is 9; the square of six exceeds the first by three nines to make 36; to this the square of nine adds five nines. The following square, moreover, namely the square of twelve, adds 7 nines to the square of nine. And thus successively the same I found for the increasing squares which are increasing by fours and for other numbers to infinity. Out of all of these come the solutions to the questions that follow.

Comments on Proposition 21

For any three consecutive odd squares, the greatest square exeeds the middle square by eight more than the middle square exceeds the least square. Let $2n + 1$, $2n + 3$, $2n + 5$ be three consecutive odd numbers

$$(2n + 5)^2 - (2n + 3)^2 = (4n + 8)(2).$$

$$(2n + 3)^2 - (2n + 1)^2 = (4n + 4)(2).$$

The excess of one difference over the other is $(8n + 16) - (8n + 8) = 8$. Or, as Leonardo puts it,

$$(gd)^2 - (bg)^2 = (bg + gd)(gd - bg) = (bd)(2).$$

$$(bg)^2 - (ab)^2 = (ab + bg)(bg - ab) = (ag)(2).$$

$$2(bd) - 2(ag) = 2(bg + gd) - 2(ab + bg) = 2(gd) - 2(ab)$$

$$= 2[(gd) - (ab)] = (2)(4) = 8.$$

Here is the principle of the difference of eights summed yielding a formula for odd squares.

$$[2(2) - 1]^2 = 1 + 8.$$

$$[2(3) - 1]^2 = 1 + 8 + (2)8.$$

$$[2(4) - 1]^2 = 1 + 8 + (2)8 + (3)8.$$

$$\vdots$$

$$[2p - 1]^2 = 1 + 8 + (2)8 + \cdots + (p - 1)8.$$

And here is a formula for even squares.

$$[2(1)]^2 = 4.$$

$$[2(2)]^2 = 4 + 12 = 4 + [4 + 8].$$

$$[2(3)]^2 = 4 + (4 + 8) + [4 + 2(8)].$$

$$\vdots$$

$$[2p]^2 = 4 + (4 + 8) + [4 + 2(8)] + \cdots + [4 + (p - 1)8].$$

$$= 4 + 3(4) + 5(4) + \cdots + (2p - 1)(4).$$

For squares which increase by threes or fours, we have these formulas.

$$(3p)^2 = 9 + 3(9) + 5(9) + \cdots + (2p - 1)(9).$$

$$(4p)^2 = 16 + 3(16) + 5(16) + \cdots + (2p - 1)(16).$$

Proposition 22

I wish to find in a given ratio the two differences among three squares.

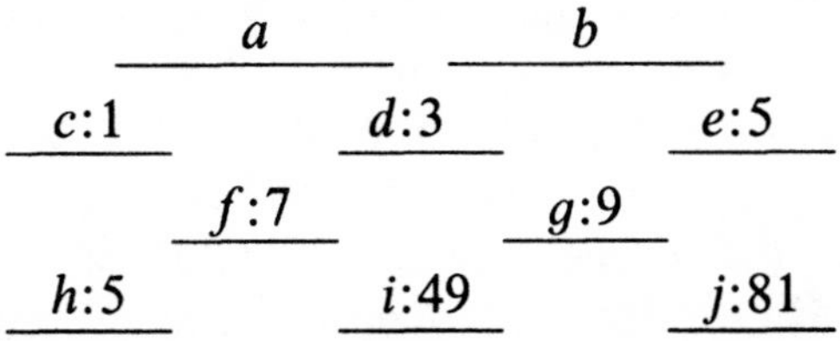

	a		b	
c:1		d:3		e:5
	f:7		g:9	
h:5		i:49		j:81

Let there be given the ratio of number $.a.$ to number $.b.$ and let the numbers $.a.$ and $.b.$ be relatively prime. The numbers $.a.$ and $.b.$ are indeed consecutive or they are not. First, let them be consecutive and let the number $.b.$ be greater than $.a.$; and let $.c.$ be the unity. From the unity $.c.$ are placed in order as many odd numbers as there are unities in the greater number $.b.$; they are $.d.$, $.e.$, $.f.$, $.g.$. The squares of the numbers $.e.$, $.f.$, $.g.$ are taken; they are the numbers $.h.$, $.i.$, $.k.$. I say that the ratio of the difference between the square $.h.$ and the square $.i.$ is to the difference between the square $.i.$ and the square $.k.$ as the number $.a.$ is to the number $.b.$, which thus is proven.

Because the unity is $.c.$ and from it are placed the consecutive odd numbers $.d.$, $.e.$, $.f.$, $.g.$; $.d.$ will be 3 and $.e.$ will be 5 and $.f.$ 7 and $.g.$ 9; and the square of the number $.e.$, namely the number $.h.$, is 25; and the square of the number $.f.$, namely the number $.i.$, is 49; and the square of the number $.g.$, namely the number $.k.$, is 81. And because the quantity of numbers $.d.$, $.e.$, $.f.$, $.g.$ is as the number of unities in the number $.b.$, and there are four of the numbers $.d.$, $.e.$, $.f.$, $.g.$, it will be manifest that $.b.$ is 4 and the number $.a.$ is 3, and manifest that the square of $.d.$, namely of three, exceeds the square of the number $.c.$ by one eight; and the square of the number $.e.$ exceeds the number $.d.$ by two eights. And the square of the number $.f.$, namely the number $.i.$, exceeds the square of the number $.e.$, that is the number $.h.$, by three eights, namely according to the number of unities that are in the number $.a.$. And also the square of the number $.g.$ exceeds the square of the number $.f.$, that is the square $.i.$, by four eights, namely according to the number of unities that are in the number $.b.$. Therefore, the ratio of the difference between the squares $.h.$ and $.i.$ to the difference between the squares $.i.$ and $.k.$ is as $.a.$ is to $.b.$, that is as 3 to 4. And if the number $.a.$ is 10 and $.b.$ is 11, according to that which is said, 10 and 11 are added together and 21, which then results, is the root of the middle square. Therefore, 19 will be the root of the smallest square and 23 will be the root of the largest square. Certainly 19 and 21 and 23 are consecutive odds; and 21 is the tenth odd number after the unity. Therefore, the square of the same 21, namely 441, exceeds the square of 19, namely 361, by ten eights. And the square of the number 23, which is the eleventh odd number after the unity, exceeds the square of the number 21, namely 529, which exceeds 441 by eleven eights. Therefore, the difference between 361 and 441, namely 80, is to the difference between 441 and 529, namely 88, as 10 is to 11. For the ratio that 80 has to 88, the same has $\frac{1}{8}$ of 80 to $\frac{1}{8}$ of 88, namely, 10 to 11. This had to be shown.

And if the numbers $.a.$ and $.b.$ are not consecutive, then they are consecutive odd numbers or they are not. Let them first be consecutive odd numbers and because the squares of even numbers increase by increasing odd multiples of four as 4, which is the

a *b*

square of two, namely the first even number which increases by the quantity of four; and 16, which is the square of the second even number, namely four, which arises from the sum of two odd multiples of four, namely 1 and 3, it will be manifest that any even square exceeds by two [more] fours the preceding even square. This is over and above the fours that this square exceeds its preceding square. This means that the third even square exceeds by five fours the second even square as the second even square exceeds by three fours the first even square, namely 16 over 4; and the fourth even square exceeds by seven fours the third even square. And thus occur all in order.

Whence we wish to find any two differences among three square numbers with ratio as two consecutive odd numbers, as we said, as 11 to 13. We shall take among consecutive even squares the middle square, which exceeds the preceding even square by 11 fours. The three squares thus can be found. If 13 is added to 11, there will be 24, of which the fourth part multiplied by 2, namely by the root of the first even square, will be 12, which is the root of the middle square; and 10 will be the root of the first square, and 14 will be the root of the third square.

We can also find the same thing between the squares of numbers increasing by threes. For example, if we wish to find two differences between three square numbers which have ratio as 19 to 21, which are consecutive odd numbers, we shall add 19 to 21, and the 40 that results we shall divide by 4; and the 10 that then results we shall multiply by 3, namely the root of the first square in the same order; there will result 30, which will be the root of the middle square. Therefore, the root of the smallest square will be 27, and the root of the larger will be 33. For the square of 30, namely 900, exceeds the square of 27, namely 729, by 19 nines; and the square of 33, namely 1089, exceeds 900 by 21 nines; and thus the ratio of the difference between 729 and 900, namely 171, is to the difference between 900 and 1089, namely 189, as 19 is to 21. This is what we wished. That which is still to be found is between the squares of numbers increasing by fours or fives or by any number.

And if the ratio of the two differences that are among the three squares are as some square $.a.$ to some square $.b.$, I shall wish to find the same three squares. I shall let the number $.c.$ be the geometric mean between $.a.$ and $.b.$. This is possible because, as is seen in Euclid, between two square numbers falls one mean number, and this number $.c.$ will

a:16 c:20 b:25 d

be produced by multiplying the root of the number $.a.$ by the root of the number $.b.$, and $.a.$ will be to $.c.$ as $.c.$ is to $.b.$; and if $.c.$ is to $.b.$ as $.b.$ is to $.d.$, then the numbers $.a.$, $.c.$, $.b.$, $.d.$ will be

e f g h

in geometric proportion. Therefore, $.a.$ will be to $.b.$ as $.c.$ to $.d.$; and let the square of the number $.a.$ be the number $.ef.$, and the square of the number $.c.$ be the number $.eg.$. Moreover, let the square of the number $.b.$ be the number $.eh.$. I say that the difference among the squares $.ef.$ and $.eg.$ and $.eh.$, namely the numbers $.fg.$ and $.gh.$, have ratio to each other the same as have the square number $.a.$ and the square number $.b.$. Which thus is proven. Because the numbers $.a.$, $.c.$, $.b.$ are in geometric proportion, the first, $.a.$, is to the third, $.b.$, as the square of the first number $.a.$ is to the square of the second number $.c.$, as is clearly demonstrated in geometry. Certainly the square of the number .a. is the number .ef., and the square of the number .c. is the number .eg.; therefore, .a. is to .b. as the number .ef. is to the number .eg.. Again, because the numbers .c., .b., .d. are in geometric proportion, as .c. is to .d., so is the square of the number .c., namely the number .eg., to the square of the number .b., namely to the number .eh.. But as .c. is to .d., so was .a. to .b.; therefore, as .a. is to .b. so is the number .e. to the number .eh.. Still, as .a. was to .b., so is .ef. to the number .eg.. Therefore, the numbers .ef. and .eg. and .eh. are in geometric proportion. And therefore, as .ef. is to .eg., so is .eg. to .eh.. Therefore, by separation, as .ef. is to .fg. so will be .eg. to .gh.. Therefore, by alternation, as .ef. is to .eg., so is .fg. to .gh.. But .ef. is to .eg. as .a. is to .b.. Therefore, as .a. is to .b. so is .fg., namely the difference between the squares .ef. and .eg., to .gh., namely the difference between the squares .eg. and .eh.. This also is to be shown with numbers.

Let, in fact, the number .a. be 16 and the number .b. be 25. Therefore, the number .e. will be 20, which is the product of the root of 16 and the root of 25. And as 16 is to 20, so is 20 to 25. And the square of the number .a., namely the number .ef., will be 256. And the square of the number .c., namely the number .eg., 400. And the square of the number .b., that is the number .eh., is 625. Whence if from .eg. is subtracted .ef., there will remain 144 for the number .fg.; and if from the square .eh. is subtracted .eg., namely 400 from 626, there will remain 225 for the number .gh.. Certainly 144 is to 225 as 16 is to 25. And this I have wished to show.

And if the ratio between the two differences lying among the three squares was not one of the cases mentioned before, namely from consecutive numbers, or consecutive odd numbers, or two squares, then we shall find solutions from a sequence of increasing multiples of eights lying among the odd squares made by consecutive numbers increasing from unity, or from a sequence of increasing multiples of fours lying among the even squares made from numbers increasing from unity by odd numbers. Let us take the ratio of the two differences lying among three square numbers to be as 2 is to 9. I shall take first the square of five, which exceeds the square of the preceding odd number by two eights. And I shall have the same, if possible, as the first square. And I shall proportion the same two eights to the eights by which the square of the odd number following five exceeds the square of five, namely three eights. And because the ratio of 2 to 3 is not as 2 to 9, to three eights I shall add four eights by which the square of nine exceeds the square of seven, and there will be seven eights. And there will be a ratio of two eights to seven eights, as 2 to 7. But the ratio of 2 to 7 is greater than the ratio which is 2 to 9. Therefore, to 7 eights I add the number of eights of the following odd square, namely that of the square of 11, of which there will be 12 eights. To that number with the two having smaller ratio than to 9, I shall double the numbers of the ratio, namely 2 and 9, and I shall have 4 and 18. And I shall consider the ratio of 4 to the following number, namely to 5, or to the two following numbers, namely to 5 and to 6, or to the three following numbers; thus, I shall find then the ratio which has 4 to 18; and this will be the case if I shall take with four the three following numbers, namely 5 and 6 and 7, which sum to 18, to which number 4 has the ratio 2 to 9. And because of this the solution to the question is found, and I shall have for the greater square the square of 15, namely 225. The number 15 is twice 7 plus one. And the square of 15 exceeds the square of 13 by seven eights. And the square of 13 exceeds the square of 11 by six eights; and the square of eleven exceeds the square of 9 by five eights; and thus the square of 15 exceeds the square of 9 by 18 eights. And the square of 9 exceeds the square of 7 by four eights; and thus the ratio of the difference which is between the square of 7 and the square of 9, namely between 49 and 81, will be to the difference between 81 and 225 as 2 to 9. And this is what I have wished to demonstrate.

All of the above questions and also similar ones in squares using doubles or triples or any numerical multiples are solved by the methods found above. For example, to the squares of 7 and 9 and 15 were applied the aforementioned questions. Therefore, if we doubled these three numbers we shall have for the root of the smallest square 14, and for the root of the middle 18, and for the root of the largest 30; and the ratio of the difference between the squares of the same numbers will be as 2 to 9. This I shall show geometrically.

Let the segment *.ab.* be 49, namely the square of seven; and *.ac.* be 81, namely the square of nine; and *.ad.* be the square

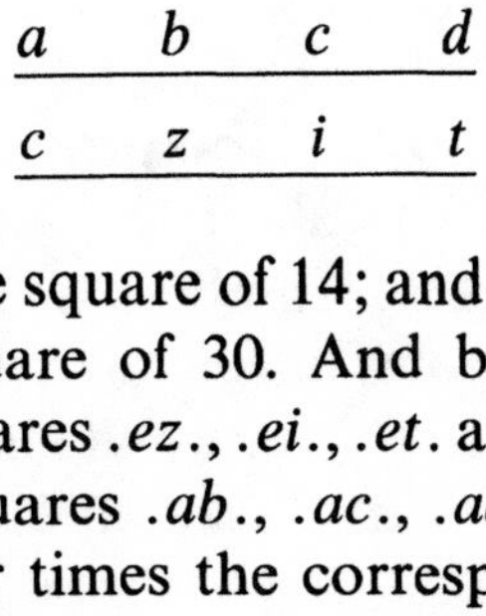

a *b* *c* *d*

c *z* *i* *t*

of 15; and *.ez.* be the square of 14; and *.ei.* the square of 18; and *.et.* be the square of 30. And because the numbers which make the squares *.ez.*, *.ei.*, *.et.* are twice the numbers which make the squares *.ab.*, *.ac.*, *.ad.*, each square *.ez.*, *.ei.*, *.et.* will be four times the corresponding one, namely *.ez.* and *.ab.*, *.ei.* and *.ac.*, *.et.* and *.ad.*. And because the total *.ei.* is four times the total *.ac.*, *.ez.* similarly four times *.ab.*, the difference *.zi.* is four times the difference *.bc.*. Similarly, it is shown *.it.* is four times *.cd.*. Therefore, *.bc.* is to *.cd.*, as *.zi.* is to *.it.*. Certainly, *.bc.* is to *.cd.* as 2 is to 9; and *.ez.*, *.ei.*, *.et.* are three times the numbers whose squares are *.ab.*, *.ac.*, *.ad.*. Each square *.ez.*, *.ei.*, *.et.* is nine times its corresponding square; therefore, the difference *.zi.* is nine times the difference *.bc.*, and the difference *.it.* nine times the difference *.cd.*; therefore, as *.bc.* is to *.cd.*, so is *.zi.* to *.it.*; and this I wished to demonstrate.

And if the ratio of the two differences which are among the three squares will be as 11 to 43, the first square will be 25, the second 729 and the third 3481. I found these in order. I first have let the middle square be the square of 23, which exceeds by eleven eights the square of 21; and I investigated the ratio that 11 has to the first following number, or to two or to more, and I did not find with these same numbers a ratio of 11 to 43, because if 12 and 13 and 14, which follow 11 in numerical order, are added, they make only 39. The number 11 has a greater ratio to that number than to 43. And if to the same 39 is added the following number, namely 15, 54 is obtained. The number 11 has a smaller ratio to that number than to 43. And because of this I doubled 11 and 43, and moreover tripled them, and made every multiple of them up to 7, and I did not find between the consecutive numbers the ratio I sought. Finally, I multipled 11 and 43 by eight and I got 88 and 344. I divided 88 by 11 and produced 8. About it I put ten more consecutive numbers and made 8 the median of them. And there were 11 numbers which summed to 88, of which the smallest number is 3 and the largest, 13. I doubled 13 and added 1, and 27 resulted, the square of which exceeds by 13 eights the square of 25. And the square of 25 exceeds by 12 eights the square of 23. And thus I found by investigation the square of 27, which exceeds the square of 5 by 88 eights, which results from a sum of numbers from 3 to 13, namely eleven numbers in order.

Next, I took 14 and its succeeding numbers up to 29, and I added them and I got 344, namely eight times 43. And to the double of 29 I added 1, and I got 59 for the root of the greatest square, namely 3481, which exceeds by 29 eights the square of the preceding odd number, namely the square of 57, of which the square exceeds by 28 eights the square of 55. And thus adding eights to the preceding squares from the square of 59 down to the square of 27, for all of them I added 344 eights, by which the square of 59 exceeds the square of 27. Therefore, the ratio of the difference between the square of five and the square of 27 is to the difference between the square of 27 and the square of 59 as 11 to 43. Also, the ratio will be found of the squares made by the double or any multiple of the found roots.

Comments on Proposition 22

Leonardo's problem here is given a, b to find x, y, z so that

$$x^2 + ta = y^2. \qquad y^2 + tb = z^2,$$

that is

$$(y^2 - x^2)/(z^2 - y^2) = ta/tb = a/b.$$

This problem is identical with problem 19 of Book II of the *Arithmetica* by Diophantus [H1, p. 151].

$$(y^2 - x^2)/(z^2 - y^2) = r.$$

Diophantus' solution takes the following form. Suppose $y = x + 1$ and $z = x + e$ with $e > 1$.

$$(x + 1)^2 - x^2 = r[(x + e)^2 - (x + 1)^2].$$

$$2x + 1 = r(2ex + e^2 - 2x - 1).$$

$$x = [e^2 - (r + 1)/r]/2[(r + 1)/r - e].$$

There are solutions for x for e between the square root of $(r + 1)/r$ and $(r + 1)/r$.

In particular, if $r = 1$, we have the problem of proposition 14. There are solutions for e between the square root of 2 and 2. Setting $e = \frac{9}{5}$ gives the solution

$$(31/10)^2 + 36/5 = (41/10)^2, (41/10)^2 + 36/5 = (49/10)^2.$$

Clearing of fractions gives the equations which we had before.

$$31^2 + 720 = 41^2, 41^2 + 720 = 49^2.$$

In case $r = \frac{1}{3}$ as it is in problem 19 of Diophantus, e must be chosen between 2 and 4. Diophantus chooses $e = 3$. Then $x = \frac{5}{2}$ and the equations are

$$(5/2)^2 + 6 = (7/2)^2, (7/2)^2 + 18 = (11/2)^2.$$

Leonardo's method, however, is different. Consistent with the theme of his book, he looks at the differences of squares and achieves results in that manner. The first case considered by Leonardo is when a and b are consecutive integers: $b = a + 1$. The solution given by Leonardo is based upon the formula of proposition 21.

$$\{[2(a+1)-1]^2 - (2a-1)^2\}/\{[2(a+2)-1]^2 - [2(a+1)-1]^2\}.$$

$$[1 + 8 + \cdots + a8] - [1 + 8 + \cdots + (a-1)8] = a8.$$

$$[1 + 8 + \cdots + (a+1)8] - [1 + 8 + \cdots + a8] = (a+1)8.$$

Therefore, the ratio of the differences of the squares is $a8/(a+1)8 = a/(a+1)$.

The example given by Leonardo uses $a = 3$, $b = 4$.

$$(2a-1)^2 = 5. \qquad [2(a+1)-1]^2 = 7. \qquad [2(a+2)-1]^2 = 9.$$

$$[7^2 - 5^2]/[9^2 - 7^2] = 24/32 = 3/4.$$

And also $a = 10$ and $b = 11$:

$$[2a-1]^2 = 19^2. \qquad [2(a+1)-1]^2 = 21^2. \qquad [2(a+2)-1]^2 = 23.$$

$$[21^2 - 19^2]/[23^2 - 21^2] = \tfrac{80}{88} = 10/11.$$

The second case considered by Leonardo is for a and b consecutive odd numbers. $a = 2t - 1$; $b = 2t + 1$. The formula from proposition 21 that is used is the following.

$$[2p]^2 = 4 + 3(4) + 5(4) + \cdots + (2p-1)4.$$

$$(2t-1)4 = [2t]^2 - [2(t-1)]^2.$$

$$(2t+1)4 = [2(t+1)]^2 - [2t]^2.$$

The ratio of the difference of the squares is then

$$(2t-1)4/(2t+1)4 = (2t-1)/(2t+1).$$

The example given by Leonardo is $2t - 1 = 11$, $2t + 1 = 13$. $t = 6$.

$$[12^2 - 10^2]/[14^2 - 12^2] = \tfrac{44}{52} = 11/13.$$

There is also a solution given by Leonardo using the formula

$$[3p]^2 = 9 + 3(9) + 5(9) + \cdots + (2p - 1)(9).$$

The example is $2t - 1 = 19$ and $2t + 1 = 21$. $t = 10$.

$$[3t]^2 - [3(t - 1)]^2 = 2(t - 1)(9).$$

$$[3(t + 1)]^2 - [3t]^2 = 2(t + 1)(9).$$

$$[30^2 - 27^2]/[33^2 - 30^2] = 19(9)/21(9) = 19/21.$$

Leonardo gives an example in which the ratio is a/b with both a and b squares. He finds the geometric mean between a and b, namely, c the product of the square roots of a and b. There are several Euclidean results on proportion used by Leonardo:

duplicate ratio ($a/c = c/b$ implies $a/b = a/c$) [H3, Book V, def. 9],
separation ($a/b = c/d$ implies $a/(b - a) = c/(d - c)$ [H3, Book V, def. 15],
alternation ($a/b = c/d$ implies $a/c = b/d$) [H3, Book V, def. 12].

This English terminology was established by the scholar of Greek mathematics, Thomas Heath.

Leonardo continues with another example in which the numerator and the denominator are neither consecutive nor squares. He considers the fraction 2/9 and searches among the different representation of 2/9 for a suitable ratio of multiples of 8: 2/9, 4/18, 6/27, 8/36, 10/45, ... He finds in 4/18 the relationship he wants: $4/18 = 4/(5 + 6 + 7)$. This gives a solution to the problem: $4(8)/[5(8) + 6(8) + 7(8)]$.

$$[4(8) + 3(8) + 2(8) + (8) + 1] - [3(8) + 2(8) + (8) + 1] = 9^2 - 7^2.$$

$$[7(8) + 6(8) + \cdots + 2(8)8 + 1)] \\ - [4(8) + 3(8) + 2(8) + 8 + 1] = 15^2 - 9^2.$$

This is again using the formula

$$[2p - 1]^2 = 1 + 8 + 2(8) + \cdots + (p - 1)8.$$

This system of searching 2/9, 4/18, 6/27, 8/36, ... terminates because $18 = 5 + 6 + 7$. Such a simple solution cannot always be found. For example, there are no solutions for 5/29, 10/58, 15/87, ...

The final technique illustrated by Leonardo is for the example 11/43. No multiple up to 8,

$$11/43, 22/86, 33/129, \ldots, 77/301, 88/344,$$

results in a solution according to his previous method. He therefore expresses 88 as the sum of eleven consecutive numbers centered at 8.

$$88 = 3 + 4 + 5 + 6 + 7 + 8 + 9 + 10 + 11 + 12 + 13.$$

Then

$$88(8) = 27^2 - 5^2.$$

Providently,

$$344(8) = (14 + 15 + \cdots + 29)(8) = 59^2 - 27^2,$$

and his problem is solved. At this point Leonardo leaves the discussion somewhat unresolved.

It is possible, however, to see that there always exist solutions to the problem although they are not, of course, unique. There exist nonnegative integers u, v, w so that for relatively prime a, b, $a < b$, we have

$$\begin{aligned} a/b = {} & [(u + 1) + (u + 2) + \cdots + (u + v)] \\ & \div [(u + v + 1) + (u + v + 2) + \cdots + (u + v + w)]. \end{aligned}$$

The equation is solved for u in terms of v and w.

$$u = [avw + aw(w + 1)/2 - bv(v + 1)2]/[bv - aw].$$

Choose positive integers v, w so that $bv - aw = 1$ using the Euclidean algorithm. avw, $aw(w + 1)/2$, $bv(v + 1)/2$ are all integers and $bv - aw = 1$; therefore, u is an integer. That u is not negative can be seen as follows.

$$\begin{aligned} 2u &= 2avw + aww + aw - bvv - bv \\ &= v(aw - bv) + (aw - bv) + avw + aww \\ &= -v - 1 + avw + aww \\ &= v(aw - 1) + (aww - 1). \end{aligned}$$

$(aw - 1)$ and $(aww - 1)$ are both positive except for the special case of $a = 1$ and $w = 1$, which is easily resolved.

We demonstrate the use of the algorithm for $a/b = 5/29$. $29v - 5w = 1$ has the Euclidean solution $v = 4$ and $w = 23$. u is then calculated from the preceding formula to be 1550. This gives

$$5/29 = [1551 + \cdots + 1554]/[1555 + \cdots + 1577].$$

The numerator is

$$[1 + 8 + 2(8) + \cdots + 1554(8)] - [1 + 8 + 2(8) + \cdots + 1550(8)] = 3109^2 - 3101^2$$

and the denominator is

$$[1 + 8 + 2(8) + \cdots + 577(8)] - [1 + 8 + 2(8) + \cdots + 1554(8)] = 3155^2 - 3109^2.$$

Proposition 23

I wish to find three squares so that the sum of the first and the second as well as all three numbers are square numbers.

I shall find first two square numbers which have sum a square number and which are relatively prime. Let there be given 9 and 16, which have sum 25, a square number. I shall take the square which is the sum of all odd numbers which are less than 25, namely the square 144, for which the root is the mean between the extremes of the same odd numbers, namely 1 and 23. From the sum of 144 and 25 results, in fact, 169, which is a square number. And thus is found three square numbers for which the sums of the first two and all three together are square numbers. If to 169 is still added the square number which is the sum of all the odd numbers from one up to 167, of which the root is 84, namely half 168, there results 7225, which is a square number and its root is

85. And thus are found four squares for which the sum of two or three or all together make a square number.

Moreover, to 7225 we can add three distinct squares and with each of them make a square number. The first is the square with root 3612 resulting from the sum of all the odd numbers smaller than 7225. The second is the square with root 720, which results from the sum of all the odd numbers which is less than a fifth part of 7225 diminished thence by the two preceding odd numbers. The third square with root 132 results, in fact, from the sum of the odd numbers which are smaller than 1/25 of 7225, diminished by twelve odd numbers of this 1/25 part. And thus can be found an infinite number of square numbers that separately or added make square numbers according to this arrangement.

Comments on Proposition 23

I wish to find three square numbers so that the sum of the first and the second as well as all three numbers are square numbers.

Leonardo uses this familiar identity to produce additional squares.

$$a + [(a - 1)]^2 = [(a + 1)/2]^2.$$

Beginning with $9 + 16 = 25$, Leonardo finds a square to add to 25 to produce another square.

$$a = 25. \qquad (a - 1)/2 = 12. \qquad (a + 1)/2 = 13.$$

$$3^2 + 4^2 = 5^2. \qquad 5^2 + 12^2 = 13^2.$$

Again, with $a = 13^2 = 169$. $(a - 1)/2 = 84$. $(a + 1)/2 = 85$.

$$13^2 + 84^2 = 85^2.$$

Again, with $a = 85^2 = 7225$. $(a - 1)/2 = 3612$. $(a + 1)/2 = 3613$.

$$85^2 + 3612^2 = 3613^2.$$

Leonardo points out that these answers are not necessarily unique by giving two other solutions.

$$8^2 + 720^2 = 725^2. \qquad 85^2 + 132^2 = 157^2.$$

Although Leonardo does not mention it, there is one more solution.

$$8^2 + 204^2 = 221^2.$$

Proposition 24

The question proposed to me by Master Theodore, Philosopher to the Emperor

I wish to to find three numbers which added together with the square of the first number make a square number. Moreover, this square, if added to the square of the second number, yields thence a square number. To this square, if the square of the third number is added, a square number similarly results.

First, three square numbers are found of which two added together make a square and the sum of the three again make a square. Let the smallest of these numbers be greater than the sum of the roots of the other two squares. Let 36 and 64 and 576 be given; the root of the second number will be 8 and the third 24. These roots are held for the second and third numbers of the three sought numbers. And I shall take for the first number some root. I shall add these three numbers together and I shall have the sum of 32 and the root. To this I shall add the square of the root and I shall have their sum of 32, the root and its square. All of this I wish to be equal to the first posed square, namely 36. I shall subtract from each part 32 and there will remain the sum of the root and its square equal to four. Whence just as in a similar procedure I shall take for the square *.ac.*

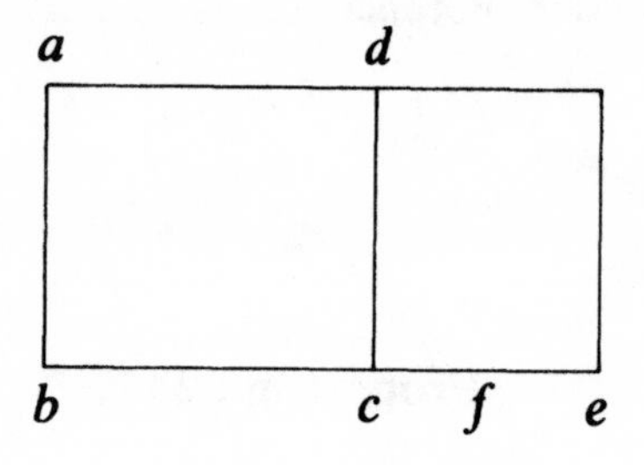

of which each side is equal to the posed root. I shall add the rectangular area *.de.*, which is the root of the square *.ac.*. Therefore, *.ce.* will be 1 and *.dc.* is the root equal to which is one of the sides of the squares *.ae.*. And I shall with *.f.* divide *.ce.* into two equal sections, *.cf.* and *.f.*, each equal to one-half unity.

And therefore we find a square and a root with sum equal to four. Manifestly, the area of the rectangle *.ae.* will be 4. The area is the product of *.ab.* and *.be.*, that is *.bc.* times *.be.*. Whence the segment *.ce.* is divided into two equal parts by *.f.* and in the direction of them is added the segment *.cb.*. The area *.bc.* times *.be.*, plus the square of the segment *.cf.* will be equal to the square of the segment *.bf.*. But the product of *.bc.* and *.be.* is 4. If the square of the number *.cf.*, which is $\frac{1}{4}$, is added to that, one will have $4\frac{1}{4}$ for the square of the number *.bf.*; but $4\frac{1}{4}$ is a number that has no root. We say that the number *.bf.* is the root of $4\frac{1}{4}$. From this, if one subtracts the number *.cf.*, which is $\frac{1}{2}$ of unity, there will remain for the root *.bc.* the root of $4\frac{1}{4}$ minus $\frac{1}{2}$, which, however, is an irrational number and will be the first sought number, and the second will be 8, the third 24.

Example: For the sum, in fact, of these three numbers one has $31\frac{1}{2}$ plus the root of $4\frac{1}{4}$. To this sum if we add the square of the first number, which is $4\frac{1}{2}$ minus the root of $4\frac{1}{4}$, one will have 36, which is a square number. To this, if is added 64, namely the second square number, there results 100, which is a square number and the root of it is 10, to which square, if is added 576, namely the third square number, one will have 676, which is a square number and the root of it is 26. And this we wished.

And to have the solution of the above stated question in rational numbers, it is shown first that when $\frac{1}{4}$ of the whole number one is added to some number which is the product of two rational numbers, one exceeding the other by unity, there results from that a square number. This is shown in the area *.ag.*, which is the product of two rational numbers,

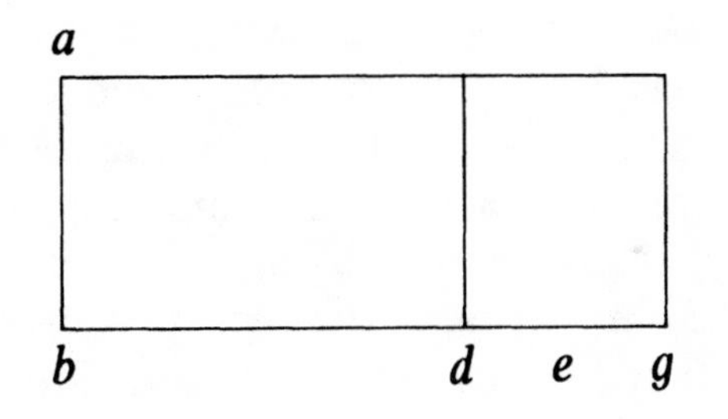

one exceeding the other by 1, which are *.ab.* and *.bg.*, and the bigger one of them is *.bg.*, and the unity *.gd.* is subtracted from the bigger *.bg.*; there will remain the number *.bd.* equal to the number *.ab.*. And the unity *.gd.* is divided in half by *.e.*. The segment *.de.* will be half the whole number one. Therefore, half *.de.* is rational. Certainly the number *.bd.* is rational. Therefore, the total *.be.* is a rational number and the square of it is rational. This square is equal to the area, *.bd.* times *.bg.*, that is *.ab.* times *.bg.*

plus the square of *.de.*. But the square of *.de.* is $\frac{1}{4}$ of unity. And the number which results from *.ab.* times *.bg.* is the product of two rational numbers, one exceeding the other by 1. Therefore, if $\frac{1}{4}$ is added to the product of the two numbers, one exceeding the other by 1, there then results a square number. This is what I wished to show.

And it is noted that all the whole numbers that are products of consecutive numbers are sums of even numbers in order. For 2, which is produced from 1 times 2, is the first even number. And 6, which is 2 times 3, is the sum of the first two even numbers. And 12, which is 3 times 4, is the sum of three even numbers, namely 2 and 4 and 6. And the sum of ten even numbers in order results in 10 times 11. The same is seen in all the remaining numbers that are products of two consecutive whole numbers. And it is known that every odd number is the sum of two consecutive numbers. Whence any odd number can be parted into two consecutive numbers, as 7 which parts into 3 and 4.

Now I wish to show that when a number of roots are taken away from a square number, and if the number of the same root is divided into two parts, one of which exceeds the other by 1, and if one of these parts is multiplied by the other, and if that which will result is added to what remains of the square after the roots were subtracted, there will result a number which is the product of two unequal numbers, the larger exceeding the smaller by 1. To demonstrate this, take the square *.ag.* and subtract the number of its roots contained in the area *.eg.*. Therefore, the number *.fg.* contains as many unities as there are roots of the square *.ag.* in the area *.eg.*. And the number *.fg.* is divided into two parts, the larger exceeding the smaller by 1, which are *.fi.* and *.ig.*, and let the larger be *.ig.*. I say that if one takes away the area *.eg.* from the square *.ag.*, what remains, namely the area *.af.* plus the area *.if.* times *.ig.*, makes a number which is the product of two unequal numbers, the greater exceeding the smaller by 1. And this is the square *.ag.* from which is taken away the area *.eg.* minus the area *.if.* times *.ig..* We put, in fact, the number *.gh.* equal to the number *.if.*, and there will remain the unity *.ih.*, which is divided into two halves, which are *.zi.*, *zh.*, and the whole *.fg.* will be divided into two equal parts by *.z.* and into two unequal parts by *.i..* Therefore, the product *.fi.* by *.ig.* plus the square of *.iz.* is equal to the square of the number *.fz..* Again, because the number *.fg.* is divided into two equal parts by *.z.*, and to *.fg.* is added the number *.bf.*, the product *.bg.* by *.bf.*, that is *.ab.* by *.bf.*, plus the square of *.fz.*, will equal the square of the number *.bz..* But the square of the number *.fz.* is equal to the area *.fi.* times *.ig.* plus the square of *.iz.* or one-half. Therefore, the products of *.ab.* by *.bf.*, that is the area *.af.*, plus the product of *.fi.* by *.ig.*, plus the square of the number *.iz.*, is equal to the square of the number *.bz..* Again, because *.ih.*, the unity, is divided into two equal parts by the point *.z.* and to *.ih.* is added the number *.bi.*, the product of *.bi.* by *.bh.* plus the square of *.iz.* will equal the square of *.bz..* But the square of *.bz.* equals the area of *.af.* plus the area *.fi.* times *.ig.* plus the square of *.iz..* Therefore, the area *.bi.* times *.bh.* plus the square of *.iz.* are equal to the area *.af.* plus *.fi.* times *.ig.* plus the square of *.iz..* The square of *.iz.* is subtracted from both and there will remain the area *.af.* plus the area *.fi.* times *.ig.* equal to the area *.bi.* times *.bh..* But the area *.bi.* by *.bh.* is the product of two numbers, one exceeding the other by 1, which are *.bi.* and *.bh..* Certainly *.ih.* is 1. Also, it will be shown with numbers.

Let, in fact, the square *.ag.* be 100, and either side will be 10. And subtract from the square *.ag.* seven roots of it minus the product *.fi.* by *.ig..* The product of the roots is the area *.eg..* There will remain 30, the area *.af.*, to which if the product *.fi.* by *.ig.* is added, that is 3 by 4, there will result 42, which is the number arising from the multiplication of *.bi.* by *.bh.*, that is 6 by 7. Certainly, the total *.bg.*

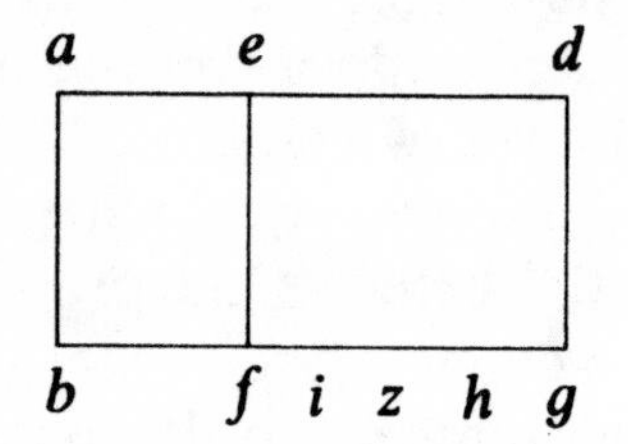

is 10, from which if is subtracted the number *.fg.*, which is 7, there will remain 3 for the number *.bf.*, to which *.fi.*, which is 3, is added; the total number *.bi.* will be 6, to which if is added the unity *.ih.*, will result in 7 for the number *.bh.*.

And after all this is demonstrated, we return to the question of the philosopher, and we proceed in the aforesaid manner. Thus, we have that the sum of the square and the root and 32 are equal to the square 36. Next, we see how many roots of 36 are in 32; this is why we divide 32 by the root of 36; there results $5\frac{1}{3}$. And because of this, we find the solutions to the aforementioned question in the given ratio of the three aforementioned squares, namely 36, 64 and 576; it is necessary that we find some square after which is subtracted $5\frac{1}{3}$ roots, there remains a number which is the product of the said unequal numbers, the larger exceeding the smaller by 1. We shall find this square if we shall pose some root number exceeding the aforementioned $5\frac{1}{3}$ roots. That we indeed can do in an infinite number of ways. Therefore, we pose arbitrarily 7 roots and we divide 7 into two parts, one exceeding the other by 1; they will be 3 and 4. And 3 is multiplied by 4 making 12. And we know, by that which is said, that when from some square is subtracted 7 roots minus 12, there will remain of this same square a number which is the product of two unequal numbers, the larger exceeding the smaller by 1. And we wish to find a square, when, after $5\frac{1}{3}$ roots are subtracted, there remains similarly a number which is the product of two numbers, one exceeding the other by 1. Therefore, $5\frac{1}{3}$ roots of the same square we seek are equal to seven roots of the said square diminished by 12. Therefore, if we add 12 to each part, there will be $5\frac{1}{3}$ roots and 12 drachmas which are equal to seven roots. We therefore take away from each part $5\frac{1}{3}$ roots; there will remain $1\frac{2}{3}$ roots which are equal to 12. Therefore, we triple both and there will be five roots equal to 36. Whence, if we shall divide 36 by 5, we shall have $7\frac{1}{5}$ for the root of the sought square, namely the first. In fact, the root of the first square was 6; therefore, proportionally as 6 is to $7\frac{1}{5}$, so are 8 and 24 to the roots of the second and third squares. But $7\frac{1}{5}$ exceeds 6 by a fifth part of itself. Therefore, if to 8 and 24 we add a fifth of them, we shall have for the root of the second square $9\frac{3}{5}$ and for the root of the third $28\frac{4}{5}$. And $9\frac{3}{5}$ will be the second of the three sought numbers, and $28\frac{4}{5}$ will be the third. And the number is yet unknown which, when added to the second and third aforementioned numbers and to the same first square, yields the square of $7\frac{1}{5}$, which is $51\ \frac{4}{5}\ \frac{1}{5}$. Therefore, we shall put for the first number a root and we shall add to it $9\frac{3}{5}$ and $28\frac{4}{5}$, and we shall have the root plus $38\frac{2}{5}$. To this we add the square of the root and we shall have the sum of the square, the root and $38\frac{2}{5}$, which will equal $51\ \frac{4}{5}\ \frac{1}{5}$ drachmas. Therefore, we take away from each part $38\frac{2}{5}$; there will remain the square plus the root equal to $13\ \frac{2}{5}\ \frac{1}{5}$ drachmas. To them we add $\frac{1}{4}$, namely the square of half a root of unity, as we did previously, and we shall have $13\ \frac{6}{10}\ \frac{9}{10}$, which is 1369 hundredths. We divide therefore the root of 369, namely 37, by the root of 100, yielding $3\frac{7}{10}$, from which we take away $\frac{1}{2}$ for half a root of unity, which will yield $3\frac{1}{5}$ for the first number. And thus this question is solved in rational numbers, and following this method the question can be solved in an infinite number of ways.

I have solved also this question in whole numbers of which the first was 35, the second 144, the third 360, and the sum of them anounts to 539, to which if added the square of the first number 35, namely 1225, there results 1764, which is a square number with root 42. To this square if is added the square of the number 144, which is 20736, there results 22500, which is a square number, and the root of it is 150. To this square if is added the square of the third number 360, namely 129600, there results 152100, which is a square number and the root of it is 390. I have found these numbers after posing these three squares, namely of 49, of 576 and of 3600, of which two and all three together add to make square numbers. And I added the roots of the second and third, namely 24 and 60; they made 84, which I divided by the root of the first square, namely by 7, and there resulted 12. And because of this I had to find a square number so that after having 12 roots taken away from it, there remainded a number made of the product of two unequal numbers, one exceeded the other by 1. Whence I took 13, and I divided it into consecutive parts, namely 6 and 7, which I multiplied together and they made 42. And I had to find a square for which 13 roots minus 42 drachmas would equal 12 roots of it. And I proceeded afterwards in the aforesaid way, and I had the aforementioned numbers. From them also I found the squares of these other three numbers, namely $10\frac{2}{3}$ and 64 and 160. And not only these three numbers can be found by this means, but also four will be found with four square numbers which sum by twos and threes and all together to square numbers. Therefore, however, with four square numbers, namely ______ and ______ and ______ and ______, I found these four numbers, of which the first is 1295, the second $4566\frac{6}{7}$, the third $11417\frac{1}{7}$, the fourth truly 79920. And their sum is 97199. To this number if is added the first square number, namely

1677025, there results 1774224, which is a square number and its root is 1332. To this square also . . .

Comments on Proposition 24

I wish to find three numbers which added together with the square of the first number make a square number. Moreover, this square, if added to the square of the second number, yields thence a square number. To this square, if the square of the third number is added, a square number similarly results.

This problem was proposed to Leonardo by Master Theodore, a philosopher at the court of Frederick II. It is likely that this problem also came from Arabic sources. It is, of course, in the tradition of Diophantus.

The following squares are found by the techniques of proposition 23, the previous theorem.

$$6^2 + 8^2 = 10^2. \qquad 10^2 + 24^2 = 26^2.$$

It is proposed by Leonardo to find quantities x, y, z and r, s so that

$$x^2 + x + y^2 + z^2 = r^2; r^2 + y^2 = s^2;$$

$$s^2 + z^2 = t^2.$$

The problem is fitted to the numerical values given in the previous proposition.

$$x^2 + x + 8 + 24 = 6^2; 6^2 + 8^2 = 10^2;$$

$$10^2 + 24^2 = 26^2.$$

Now he needs only to find x so that $x^2 + x + 32 = 6^2$. This yields for x the solution of the equation $x^2 + x = 4$. He then shows

$$x^2 + x + 1/4 = 17/4.$$

$$(x + 1/2)^2 = 17/4.$$

$$x = (1/2)(-1 + \sqrt{17}).$$

In some of the passages that follow, Leonardo uses the word *drachma*, which is a small Greek coin, as a synonym for the unity.

Leonardo also writes some fractions in a way not in use today.

51 4/5 1/5 means $51 + 4/5 + 1/25$.

13 2/5 1/5 means $13 + 2/5 + 1/25$.

13 6/10 9/10 means $13 + 6/10 + 9/100$.

Leonardo now wishes to find a rational solution to the previous problem. He builds such a rational solution by considering some multiple of the 6, 8, 10, 24, 26 solution. He proposes then to solve the following system for x and k so that both are rational.

$$x + 8k + 24k + x^2 = (6k)^2; (6k)^2 + (8k)^2 = (10k)^2;$$
$$(10k)^2 + (24k)^2 = (26k)^2.$$

The equation $x^2 + x + 32k = (6k)^2$ can be rewritten as

$$x^2 + x + (16/3)R = R^2 \quad \text{or} \quad x^2 + x = R^2 - (16/3)R,$$

where $R = 6k$. Finding rational solutions to this equation is again like problems found in the *Arithmetica* of Diophantus, such as problems 20 and 22 of Book II [H1, pp. 151, 152]. Such problems also appear in the *Fakhri* of Alkarkhi [Wo, p. 73]. The technique used is to set $x = R - a$ for some a that will eliminate the quadratic terms and give rational solutions for x and R. This is the essence of Leonardo's method. The substitution $x = R - a$ yields the equation

$$(R - a)(R - a + 1) + (16/3)R = R^2.$$

This yields, solving for R, $R = 3a(a - 1)/(6a - 19)$, and also,

$$x = a(3a - 16)/(19 - 6a).$$

For $19/6 < a < 16/3$, both x and R will be positive. Leonardo's values come from $a = 4$, the first acceptable integral value. They are $a = 4$, $R = 36/5$, $x = 16/5$, $y = 48/5$, $z = 144/5$. Other values of a will give other solutions.

Leonardo also finds an integral solution to the problem. Beginning with the setup

$$x + 24 + 60 + x^2 = 7^2; 7^2 + 24^2 = 25^2;$$
$$25^2 + 60^2 = 65^2.$$

This equation,

$$x^2 + x + 84 = 7^2,$$

can be written as

$$x^2 + x + 12(7) = 7^2.$$

Letting 7 be generalized to R leads to the equation

$$x^2 + x + 12R = R^2.$$

The substitution $x = R - a$ leads to the equation

$$(R - a)(R - a + 1) + 12R = R^2.$$

This has rational solutions $R = a(a - 1)/(2a - 13)$. $a = 7$ gives Leonardo's solution, $R = 42$, which leads to a complete integral solution: $x = 35$, $y = 144$, $z = 360$ as Leonardo describes.

Leonardo next generalizes the problem further by increasing the number of unknowns. The generalized problem is

$$w^2 + w + x + y + z = r^2.$$

$$r^2 + x^2 = s^2.$$

$$s^2 + y^2 = t^2.$$

$$t^2 + z^2 = u^2.$$

The blanks left in the manuscript seem to leave it to the reader to supply the missing numbers from the given hints. It is also possible that Leonardo intended to continue his paper; it does end abruptly.

Here is a construction that supplies the missing numbers and is almost certainly, as the reader can judge for himself, Leonardo's solution. First Leonardo selected three continuing Pythagorean triples.

$$7^2 + 24^2 = 25^2. \qquad 25^2 + 60^2 = 65^2. \qquad 65^2 + 420^2 = 425^2.$$

The second two triples are simply multiples of the triples 5, 12, 13 and 13, 84, 85, which he used before. The equations then are fitted to some multiple of these three triples to give a rational solution just as before with one fewer variables.

$$(7k)^2 + (24k)^2 = (25k)^2.$$

$$(25k)^2 + (60k)^2 = (65k)^2.$$

$$(65k)^2 + (420k)^2 = (425k)^2.$$

Set $x = 24k$, $y = 60k$, $z = 420k$. A solution will be obtained provided

$$w^2 + w + 24k + 6k + 420k = (7k)^2.$$

This equation simplifies to

$$w^2 + w + 504k = (7k)^2.$$

Further simplify by letting $R = 7k$. This yields the Diophantine equation

$$w^2 + w + 72R = R^2.$$

Rational solutions to the equation are given by using the substitution $w = R - a$. This is again Leonardo's method.

$$(R - a)(R - a + 1) + 72R = R^2.$$

This equation has solution

$$R = a(a - 1)/(2a - 73).$$

This gives

$$w = a(72 - a)/(2a - 73).$$

Positive values for R and w are obtained provided $73/2 < a < 72$. The first permissible integer in the interval is 37. For $a = 37$ one has Leonardo's solution $R = 1332$, $w = 1295$, $k = R/7 = 1332/7$, $x = 31968/7$, $y = 79920/7$, $z = 79920$.

$$1295^2 + 1295 + 31968/7 + 79920/7 + 79920 = 1332^2.$$

$$1332^2 + (31968/7)^2 = (33300/7)^2.$$

$$(33300/7)^2 + (79920/7)^2 = (86580/7)^2.$$

$$(86580/7)^2 + 79920^2 = (566100/7)^2.$$

The missing numbers for the blanks left by Leonardo are therefore 49, the square of 7, 576, the square of 24, 3600, the square of 60, and 17640, the square of 420. Of course, since Leonardo gave the solution to the problem instead of the given squares, it is necessary to work the problem backwards to obtain the Pythagorean triples he used. Certainly, Leonardo could have begun with some other multiples of the relatively prime 7, 24, 60, and 420, but that seems unlikely.

REFERENCES

[A] Anbouba, A. "Un traite d'Abu Ja far [al-Khazin] sur les triangles rectangles numerique." *Journal for the History of Arabic Science* 3 (1979): 134–178.

[B] Boncompagni, Baldassarre. *Liber quadratorum.* In *Scritti di Leonardo Pisano.* Rome, 1862.

[D] Dickson, L. E. *History of the Theory of Numbers.* Vol. 2. Washington D.C.: Carnegie, 1920.

[Go] Goitein, S. D. *A Mediterranean Society.* Vol. 1. Berkeley, Calif.: Univ. of California Press, 1967.

[Gr] Grant, Edward. *A Source Book in Medieval Science.* Cambridge, Mass.: Harvard University Press, 1974. Leonardo Pisano, The Book of Squares, pp. 114–129.

[H1] Heath, Thomas L. *Diophantus of Alexandria. A Study in the History of Greek Algebra.* 2nd ed. [The *Arithmetica* of Diophantus] Cambridge: Cambridge University Press, 1910.

[H2] Heath, Thomas L. *A History of Greek Mathematicians.* Vol. 1. New York: Dover, 1921, 1981.

[H3] Heath, Thomas L. *The Thirteen Books of Euclid's Elements,* 3 Vols. New York: Dover, 1921, 1956.

[K] Kantorwicz, E. *Kaiser Friedrich der Zweite.* Düsseldorf: Kupper-Bondi, 1963. Authorized English version, *Frederick the Second.* Translated by E. O. Lorimer. New York: Richard R. Smith, 1931.

[L] Loria, Gino. "Leonardo Fibonacci" In *Gli Scienziati Italiana*, pp. 4-12. Rome: Aldo Mieli, 1923.

[Mc] McClenon, R. B. "Leonardo of Pisa and his Liber Quadratorum." *American Mathematical Monthly* 26 (1919): 1-8.

[M] ben Musa, Mohammed, al-Khowarezm. *Calculating by Completion and Reduction.* Translated by Frederic Rosen. London, 1831.

[O] Ore, Oystein. *Number Theory and Its History.* New York: McGraw-Hill, 1948.

[P1] Picutti, Ettore. "Il *Libro dei Quadrati* di Leonardo Pisano." *Physics* 21 (1979): 195-339.

[P2] Picutti, Ettore. "Leonardo Pisano." *Le Scienze* no. 164 (1982); *Le Scienze, Quaderni* (1984): 30-39.

[S] Sesiano, Jacques. *Books IV to VII of Diophantus' Arithmetica.* Translation into English. New York: Springer Verlag, 1982.

[Ve] Ver Eecke, Paul. Leonard de Pise. *Le livre de nombres carrés.* French translation of The Book of Squares by Leonardo. Paris: Blanchard, 1952.

[Vo] Vogel, Kurt. "Fibonacci, Leonardo or Leonardo of Pisa." In *Dictionary of Scientific Biography.* Vol. 4. New York: Chas. Scribner's Sons (1970), pp. 604-613.

[We] Weil, André. *Number Theory: An Approach Through History from Hammurapi to Legendre.* Boston: Birkhäuser, 1983.

[Wo] Woepcke, Franz. *Extract du Fakhri. Traité d'algèbre par Abou Bekr Mohammed Ben Alhacan Alkarkhi.* New York: George Olms, 1853, 1982.

INDEX OF PERSONS

INDEX OF TERMS

I

L

N

P

R

S

T

V